PRINCIPLES OF
Cattle Production

Moulton
College
NORTHAMPTONSHIRE

Profit through Skill

PRINCIPLES OF
Cattle Production

C.J.C. PHILLIPS

BSc, MA, PhD

Head, Farm Animal Epidemiology and Informatics Unit
Department of Clinical Veterinary Medicine
University of Cambridge
UK

CABI *Publishing*

CABI *Publishing* is a division of CAB *International*

CABI Publishing
CAB International
Wallingford
Oxon OX10 8DE
UK

Tel: +44 (0)1491 832111
Fax: +44 (0)1491 833508
Email: cabi@cabi.org
Web site: http://www.cabi.org

CABI Publishing
10 E 40th Street
Suite 3203
New York, NY 10016
USA

Tel: +1 212 481 7018
Fax: +1 212 686 7993
Email: cabi-nao@cabi.org

A catalogue record for this book is available from the British Library, London, UK.

Library of Congress Cataloging-in-Publication Data
Phillips, C.J.C.
    Principles of cattle production / C.J.C. Phillips.
      p. cm.
    Includes bibliographical references (p.).
    ISBN 0-85199-438-5 (alk. paper)
    1. Cattle.   I. Title.
  SF201 .P48 2001
  636.2′ 1--dc21   00-058601

ISBN 0 85199 438 5

Typeset by AMA DataSet Ltd
Printed and bound in the UK by Biddles Ltd, Guildford and King's Lynn

Contents

Preface

Cattle are the main farm animal that is used for meat and milk production for human consumption, providing about 18% of protein intake and 9% of energy intake. Yet despite their obvious value in feeding the human population, cattle farming systems are attacked by members of the public for creating possible health risks, for providing inadequate attention to animal welfare and for alleged adverse effects on the environment. This book describes the scientific principles of cattle production and critically considers the strengths and weaknesses of the latest methods of farming dairy and beef cattle. It is particularly directed at students of agriculture, animal science and welfare and veterinary medicine, cattle husbandry advisers and leading farmers.

Farming methods that provide for optimum welfare of cattle are considered in detail. The basic requirements for housing and an adequate environment for cattle are described, as well as problems that cattle encounter in unsuitable accommodation. Some of the major cattle diseases are described individually, with attention given to those causing major loss of profitability, in particular mastitis and lameness, and examples of new diseases that have had a significant impact, such as bovine spongiform encephalopathy (BSE). The metabolic diseases are considered mainly in relation to high-producing dairy herds, and essential elements of prophylaxis are discussed.

Cattle nutrition is principally considered in relation to feeding practices in temperate zones, where food accounts for most of the farm expenditure. Most attention is directed to the high-producing dairy cow, and the relations between feeding management, milk quality and production diseases are described. The different systems for feeding beef cattle are also considered, and grazing systems examined for both beef and dairy cattle.

The book covers the principles of cattle reproduction, describing the latest techniques for breed improvement as well as the reproductive technologies that can be used to achieve these improvements. The merits of the different breeds for dairy and beef production are discussed. Oestrous behaviour in the cow is given special consideration in view of its importance to reproductive management.

As well as describing the latest methods of cattle farming, the book pays particular attention to the impact of cattle farming on the environment. In the light of this, the future roles for cattle are considered in relation to the needs of both developed and developing regions of the world. The evolution of cattle production systems in different parts of the world is also described to place this in context.

It is hoped that the book will be useful for all those involved in the cattle farming industry, enabling them to develop systems that meet modern requirements for safe food, produced in accordance with the animals' welfare requirements and with minimal or no adverse impact on the environment. I owe a debt of gratitude to all those that have helped me to develop my interest in cattle over the years – David Leaver, who gave me inspiration to learn more about cattle nutrition and behaviour; John Owen, who encouraged and helped me to develop a research programme with dairy cows at Bangor; many research students and associates, particularly Paul Chiy, who spent tireless hours helping me in cattle research and my wife, Alison, for her support throughout. I am also grateful to Insight Books, Farming Press Books Ltd and the *Journal of Dairy Science* for permission to reproduce figures, and to my father, Michael Phillips, for reading and commenting on the first chapter.

Clive Phillips

The Development of the World's Cattle Production Systems

Historical Development

Cattle evolved in the Indian sub-continent and only spread to other parts of Asia, northern Africa and Europe after the Great Ice Age, about 250,000 years ago. Two distinct subtypes are distinguishable – the humped *Bos primigenius namadicus*, the forebear of today's zebu cattle, and *Bos primigenius primigenius*, which had no hump and gave rise to modern European cattle. These wild cattle, or aurochs, were large with big horns and powerful fore-quarters compared to domesticated cattle. The bulls were usually dark brown to black, and the cows, which were much smaller than the bulls, were red-brown. Even in prehistoric times man clearly had a close association with cattle. Cave paintings in Europe show the aurochs both running wild on grassland and being preyed upon by men with arrows and spears. Their carcasses provided not only meat but valuable hides for tents, boats and clothing and bones for fishhooks and spears.

The last aurochs were killed on a hunting reserve in Poland in 1627. Domesticated cattle were developed from wild cattle (*Bos primigenius*) in the Middle East, probably about 8000–10,000 years ago. The reason for domesti-cating the aurochs is not clear. Early domesticated cattle were undoubtedly used for the production of milk, meat and for draught power, but even as early as the Stone Age cattle also had a dominant role in religion. This mainly related to their power–fertility symbolism, which derives from their strength, aggression and the ability of bulls to serve large numbers of cows. The bull came to dominate the religions of the Middle East and North Africa in particular. The ancient Egyptians worshipped the bull god, Apis, which was embodied in bulls that were selected from local herds. They were ritually slaughtered at the end of each year, after which they were embalmed and ceremoniously placed in a tomb in Saqqarah. The ancient Egyptians also worshipped cow goddesses, which represented fertility and nurture. Signifi-cantly, in Hebrew culture, as the people changed from being a warrior people to being farmers, the image of the bull changed from aggression to virility.

The spread of cattle farming across Asia and Europe was as much caused by the invasions of nomadic herdsman from the Eurasian steppes as the Middle Eastern influence. These invasions started as long ago as 4000 BC, when the European Neolithic farmers were conquered by the herdsmen on horseback who brought traditions of raising cattle on the steppes. These farmers were settled agriculturists, keeping small numbers of livestock and growing cereals. Security was provided by investing in the land, returning nutrients to build up fertility and trading peacefully between small communities.

Cattle had a crucial role in both religion, principally for sacrifice, and as a tradable commodity. In many European countries the word for 'cattle' is synonymous with 'capital'. The resistance of the people of the Italian peninsular to encroachment from Rome was fought under a banner of their cattle culture, the word Italy meaning 'the land of cattle'. When the people from the Asian steppes invaded they brought few cultural advances but a new warrior-like attitude, where security was valued as well as the ability to move fast (on horseback), with little allegiance to a particular place. Warriors were expected to expropriate cattle, often for sacrifice to appease the gods. The influence of these warriors was particularly pronounced in the west of Europe, where the Celtic descendants of the Eurasian herdsman developed a powerful cattle-based culture. Some historians believe this fuelled the colonizing tendencies of the Iberian and British peoples in particular.

The warriors from the Asian steppes also migrated into India, where the cow acquired a major religious significance. Here the population density was low and large areas were forested before domesticated cattle were widely kept. As the population grew, an increase in crop production became inextricably linked with the use of cattle for tillage. It became impossible for everybody to consume beef as the animals were required for draught purposes, and the cows were required to produce offspring to till the soil. The consumption of beef became restricted to the upper classes, in particular the Brahmin sect, and a strict class system evolved. When increased population further restricted the use of cattle for beef consumption, strict regulations were introduced that prevented beef consumption altogether.

Nowhere exemplifies the problems facing cattle production systems in developing countries better than India. With one of the highest cattle population *per capita* in the world, this vast country has had to cope with increased human population pressure and the requirement to maintain inefficient cattle production systems for religious reasons. Nowadays, cattle in India have assumed the role of scavengers and they compete less with humans for food resources, as less than 20% of their food is suitable for humans. Most is either a by-product of the human food industry or is grown on land that cannot be used to produce human food. They have become an essential and valuable part of the agrarian economy, but two problems remain. First, the inability to slaughter cows leads to the maintenance of sick and ailing animals, although some are sold to Muslims, for whom slaughter is not against their religious beliefs. Second, the increased livestock population has led to overgrazing of some grassland areas, which were first created when India's extensive forests were felled. The cultivable land area has been declining by over 1% per year and at the same time the livestock population has increased considerably, by 56% between 1961 and 1992.

Some of the grazing areas used for cattle *could* be used for the production of human food, but because of the high social status accorded to those with large herds, the Indians are turning to grassland improvement to support their expanded herds. Water retention properties of the land are improved by contour ploughing and trenching. Fertilizer nitrogen and phosphorus are used in greater quantities. In some areas sustainable use of grassland resources is encouraged by the incorporation of legumes into the sward, which can contribute substantial quantities of nitrogen. This is best achieved by inter-cropping to improve water and mineral resource use. Not all intercropping systems are better than monocropping; in extreme drought the intercrop can compete for resources to the detriment of the main crop.

In many areas of the world, the fencing of grazing resources has developed as an important measure to control the movement of stock. It is no guarantee that overgrazing will not occur and it does not create any extra land, but it is an effective management tool to allow agriculturists to use the available food resources most efficiently. The controlled burning of weeds is another management tool to allow productive species to be introduced. Fodder banks allow soil reserves to accumulate and fodder supplies to match ruminant numbers. However, until there is greater control over stock numbers and the management of grazing, resources will continue to be overused, with the resulting deterioration of production potential.

In Spain the ideological significance of cattle is deeply rooted in the culture, brought by the Celtic invasion initially and later by the Romans. The bullfight signifies the trial of strength between man and the forces of nature. The consumption of beef reared on the Spanish plains has always been popular but for a long time the warm climate meant that spices had to be added to meat as it spoiled rapidly. When Christopher Columbus set off to find a quick route to the East, he found something of much greater significance for the cattle industry. The virgin territory of the New World provided pastures for rearing cattle of superior quality to the arid interior of Spain and paved the way for colonization of most of the Americas. With no natural predators, the Longhorn cattle rapidly multiplied, and by 1870 there were over 13 million cattle on the Argentinian pampas alone. The principal South American exports were salted beef and cattle hides. In the late 19th century refrigerated transport enabled carcasses to be sent to Europe to fulfil the rising demand for beef. Most of the production was, and in places still is, on large ranches or haciendas, so that the production system and the profits were in the control of a few families. This oligopoly of agricultural production in the Iberian peninsular and their colonies prompted regular revolts by the peasants that are reminiscent of those in Europe in the Middle Ages, most recently in Portugal in the 1970s. The most recent South American revolt emanated at least in part from poverty of the farm workers, or *campesinos*, in Chile in the early 1970s.

Another large scale colonization with beef cattle, that of North America, began with the industrial revolution providing wealth for a new British middle class, who were able to afford beef on a regular basis. The English aristocracy had in the Middle Ages gained a reputation for excessive feasting on a variety of meats, with beef being the most favoured. The *nouveau riche* of the 19th century required choice joints to feed their large families, and English breeders selected smaller, better formed cattle than the Longhorn that was by this time common in South America. Breeds such as the Hereford were developed, which could be fattened in two grazing seasons, whereas the larger animals might require up to 3 years. A key figure in the development of British breeds was Robert Bakewell, who particularly selected cattle for meat production, rather than the dual purposes of meat and milk production.

In the late 19th century British and American pioneers began to search for new pastures for cattle to provide for the growing demand for beef in Europe in particular. The western ranges that covered much of the US interior were home to about four million buffalo that had roamed free for about 15,000 years. In a 10-year period, from 1865 to 1875, the Americans and several European 'game hunters' systematically slaughtered the buffalo, mainly for their hides, which were more highly prized than cattle hides because of their greater elasticity. Coincidentally, perhaps, the slaughter of the buffalo greatly assisted in the subjugation of the indigenous Indians, who, deprived of their livelihood, became dependent on the colonizers for resources. Many assisted in the buffalo slaughter and then turned to subsistence farming in the reservations. A rangeland management system that had been sustained by the Indians for several thousand years had been destroyed almost overnight. The system that replaced it was funded by investment from abroad, especially from Britain, which provided funds to purchase cattle stock, the expansion of the railways and later the development of refrigerated transport. The occupation of the rangeland by cattle ranchers was facilitated by one simple invention, barbed wire, which could be used by the 'cowboys' to stake a claim to as much land as each felt he could manage. Publicly owned rangeland was, and still is, leased for a sum well below the market value.

The USA grew in stature as a world power as Britain declined, and with the increase in affluence came the demand for well-fattened beef for home consumption. Then, instead of the cattle being finished on the range, they began to be transported for fattening on cereal-based diets in feedlots of the southern, one-time confederate states.

The Demise of Pastoral Nomadism

In some parts of Africa, nomadic systems of keeping cattle have been largely maintained at a time when they were unprofitable and politically difficult to sustain in other parts of the world. Their prevalence in Africa is largely as a result of the prevailing geographical conditions in tropical and sub-tropical areas. In the equatorial region, luxuriant plant growth makes it difficult for grass to compete with taller, more productive crops and tropical diseases are detrimental to the keeping of cattle. Here an intolerance to milk lactose in the native African people makes milk and milk products difficult to digest, and there are problems of preserving milk products in warm, humid conditions.

North and south of the equatorial belt there exists an area of less intensive agriculture, caused largely by the low rainfall. Traditionally inhabited mainly by indigenous African game, which are better adapted to the conditions but not as suitable for domestication, this region has for several hundred and, in some areas, thousands of years been the preserve of nomadic cattle keepers, such as the Masai of the Great Rift Valley of Kenya and Tanzania. The availability of grazing varies with region and with season so nomadic systems evolved, whereby the cattle herders move their stock in set patterns to find pasture land that will support their animals. Being nomadic, the herders have few possessions and cattle, like other property, are communally managed in the tribal groups. The balance between food availability and stock numbers has traditionally been managed by village councils, whose prime consideration is to maintain the animals in a healthy, productive state. They do this by attempting to prevent a shortage of grazing, which would result in the animals declining in productivity. In extreme cases it has led to tribal wars, resulting in the killing of many cattle and some humans, thus restoring the population balance. Nowadays, the village councils are sometimes dismissed in attempts to introduce a market-led economy, and the subsequent exhaustion of the grazing resources leads in the long term to reduced productivity.

For many in Africa, cattle act as the prime source of security. They provide meat, milk and blood for food, dung which can be dried and burnt for fuel, and hides and other parts of the body for a variety of uses. Those who do not have their own cattle can usually share in the benefits that they provide. The cattle have additional advantages as a store of wealth of being mobile and naturally able to regenerate, which means that the population can expand and contract according to the prevailing conditions. Money would be of much less value. Such a delicate balance between nature, man and his domesticated animals survived many centuries, but is now increasingly under threat from the forces of change that are bringing Africa into line with the developed world. The ideology of self-advancement espoused by capitalism stands in marked contrast to the communal ownership of cattle by the nomadic tribesmen. Colonial occupiers often did not understand the system and attempted to confine the nomads to certain areas, to prevent tribal warfare and to introduce western farming methods. When overgrazing resulted, they attempted artificially to match stock numbers to land availability and encouraged the nomads to settle and grow crops. However, the greatest damage done by the colonizers was to instil materialistic desires in the hearts of the African people and to believe that their own living standards could be attained in Africa by pursuing European farming and managerial techniques. As with the bison in North America, a system in perfect balance was destroyed, not quite as rapidly and not as completely, but the consequences for the continent may yet prove catastrophic.

More recently, the increase in the populations of both man and his domestic animals, largely as a result of the introduction of modern medicines to prolong life, has increased pressure for the best land to be used for cropping rather than grazing. This has intensified the overgrazing problems and further marginalized the pastoral nomads. South of the equatorial belt there has been more emphasis on introducing cattle 'ranches', with some success. However, this and other semi-intensive stock-raising methods rely on producing a saleable product, mostly to the world market because of the inability of the local people to pay for a commodity which is relatively expensive to produce. Many developed countries have erected barriers to meat imports to protect their own markets, and sometimes to protect themselves against the introduction of disease. As soon as more intensive methods are used to produce meat for the world market, the cost of inputs, many of which are taken for granted in the West, increases out of proportion. Concentrate feeds, veterinary medicines, managers trained in intensive cattle farming, all of which are much more expensive in Africa relative to meat price than in developed countries, necessitate that the products are sold on the world market rather than locally.

Similar nomadic systems have evolved in marginal areas in other regions of the world, but not on the scale of those in Africa. Where land is more productive, settled farming has over the last 2000 years or so replaced nomadism, but small migrations persist. These may even operate within a farm. In mountainous regions of Europe, such as the Alps and regions of North Wales, farmers may own a lowland region for winter grazing and have grazing rights in the mountains for the summer grazing. Formerly cattle were moved on foot by the stockmen between the two, but nowadays motorized transport is usually employed.

The Growth of Dairy Production Systems

For most of the second millennium AD, milk was produced for home consumption in villages, and cows were kept in the cities to produce milk for the urban populations. A rapid expansion of dairy farming in industrialized regions can be traced back to the advent of the railway. In Britain, for example, it meant that milk could be transported from the west of the country to the big cities, especially London, Bristol and the urban centres in the north. Nowadays transporting milk and milk products is largely by road, but the centres of dairying remain in the west, where the rainfall is high and there is a plentiful supply of grass for much of the year.

There exists in many developing countries a continued migration from rural to urban areas. Despite land resources being usually adequate for food production in rural areas, in the cities many rural migrants have inadequate supplies of good quality food, because they cannot afford it and the deterioration in foods that occurs when food is transported from the countryside. This is particularly evident for milk and dairy products, which are vital for infants as a source of minerals, particularly calcium, vitamin A and highly digestible energy and protein. The rapid deterioration of milk and dairy products in the warm conditions prevailing in sub-Saharan Africa has encouraged the establishment of some small farms in the cities, but also the establishment of suburban farms, where supplies can be brought to the farm and milk transported to the city in a short space of time. The biggest problem for these farms, which often have limited land, is to secure adequate forage resources for the cows. Distances to rural areas are often too long for the import of large quantities of fresh fodder, and conserved fodder may be inadequate for the maintenance of cattle in the rural areas, as well as being expensive and bulky to transport. This can result in conflict between the settled agriculturists and the migrant pastoralists in the rural areas. Of increasing interest is the use of by-products, such as paper and vegetable wastes, in the suburban dairy production systems. These non-conventional by-products are beginning to be used with benefits to the environment and the efficiency of land use.

Cities are not just a centre of human population, but also of industrial development, and the recent growth of urban industry has left the problem of waste disposal. Some wastes, e.g. from the food and drink industry, are used without modification for cattle production. However, the wastes are characterized by variable nutritional value and poor hygienic quality and are more suited to feeding to ruminants than monogastric animals. Brewers' and distillers' grains are perhaps the most often utilized by-products. Many other wastes do not have an established outlet and, therefore, cost money to be safely disposed of, alternatively they may create a public health hazard if they are disposed of carelessly. Some can be utilized for cattle feed, but others contain toxic agents such as arsenicals in waste newspaper or a variety of transmissible diseases. Zoonoses are of particular concern, especially since the recent transmission of a spongiform encephalopathy from animal carcasses to cattle and thence to humans in the UK. Some feel such recycling practices are unethical, but they predominate in nature and are in the interests of the development of an efficient industry. It is therefore not surprising that international bodies such as the Food and Agriculture Organization of the United Nations (FAO) and the World Bank have identified peri-urban dairying as showing the highest potential for meeting the nutrient need of urban consumers.

Cattle Production Systems and Climate

Cattle are now kept in all the major climatic regions, which demonstrates the importance of this herbivore as the major animal to be domesticated for the provision of food. As a result of the large amount of heat produced by the microbial fermentation of coarse grasses and their large size, they thrive better than most other domesticated animals in cold climates. The provision of a naturally ventilated shelter enables cattle to be kept for milk production in places such as Canada, where winter temperatures fall well below 0°C. Cattle for meat production can be kept without shelter at sub-zero temperatures, but feed intake will be increased to provide for maintenance below the lower critical temperature of the animals. Breeding cows under such conditions are usually of the more endomorphic type, such as the Hereford.

Despite their successful integration into farming systems in extreme climates, cattle are best kept in temperate environments where a steady rainfall enables grass to grow for much of the year. In some parts of the Southern Hemisphere, such as New Zealand and southern Chile, grass will grow all year and grazing systems predominate. In Britain the colder conditions in winter means that most cattle are housed for about 6 months of the year. Mediterranean climates are often too dry for cattle and the keeping of sheep and goats is more common. Because of their low intake requirements they can survive on sparser vegetation better than cattle, and sheep in particular can survive with less water than cattle, producing a faecal pellet that is harder and drier. Mediterranean cattle production systems are therefore more likely to be intensive, relying on crops such as forage maize rather than grazing, as in the Po valley of Italy.

At high temperatures cattle reduce their production levels unless they are given shade, cooling and a highly concentrated diet to minimize the heat increment of digestion. For cattle not provided with these modifications, their morphology adapts so that they absorb less heat (see Chapter 5) and lose it more readily. In some hot regions, cattle have become well adapted to their environment and traditional systems have persisted. In many developing countries, the cost of modifying the environment is prohibitive, yet small-scale cattle herding has over the last few centuries replaced hunter–gatherer societies as the more reliable form of subsistence agriculture. As mentioned, cattle provide nutrition in the form of meat and offal, milk and occasionally blood, clothing from leather and for settled agriculturists a means of tilling the land. In many traditional societies, such as the Nuer of the southern Sudan, cattle adopt a central role in the functioning of the society. The size of a cattle herd indicates a herdsman's status, and cattle may be used as a form of currency for major transactions, e.g. marriage dowry, where the bulls also become a fertility symbol.

In many parts of the developed and developing world cattle production systems have intensified this century. Average herd sizes have increased by a process of amalgamation of small units and an increase in purchased feed use. Only when dramatic political changes in eastern Europe disrupted the agricultural infrastructure in the 1990s, did herd sizes decline, as land was returned to those that had owned it before it was seized by Communists in the first half of this century. However, the economies of scale have in some of the more liberal ex-Communist countries rapidly increased herd size in private ownership, especially since the descendants of former owners often did not have the skills to profitably farm the small land areas returned to them.

The intensification process in other parts of the world has culminated in the development of cattle feedlots, with several thousand animals in a unit. This parallels the intensification of poultry and pig farming, but is less widespread.

Modern Trends

The end of the second millennium AD brought an increased suspicion by the public of farming practices, which are attacked for producing food that is not safe to eat, in a manner that damages the environment and is inhumane to the animals. This is partly in response to the intensification of modern farming methods, induced by the move to company ownership of farms and the introduction of new technologies that are directed at increasing the profitability of cattle farming. It has enabled, for example, milk yield per cow in the UK to increase from 3750 l per cow in 1970 to 5790 l in 1998. At the same time the mean herd size has increased from 30 to 72 cows, with the annual milk sales per producer increasing from 112,500 to 416,880 l over the same period.

The issue of food safety stems partly from the removal of the farming process from the control of the public; indeed many have no knowledge of how food is produced. Members of the public are less willing to accept a risk if they have no control over it. In industrialized countries the prosperity that has been generated since the Second World War has enabled more people to eat outside the home, and they therefore lose control over the cooking of the food as well as its production. People are inclined to spend more on processed food; indeed this is often demanded by their busy life schedules. As stated, they are concerned at the intensity of modern farming methods, which may reduce species diversity on farms and pollute the environment, thereby endangering safety. These concerns and others have led to many consumers opting to buy food products that come from farmers who can demonstrate more responsibility.

Perhaps the most successful scheme that farmers can join to demonstrate this attention to sustainability is the organic farming movement, often known as ecological farming. This movement is characterized by the systems for producing cattle being environmentally and socially sustainable and using a minimum of artificial inputs. As much as possible, organic farming fosters the use of crop rotations, crop residues, animal manure, legumes, green manure, off-farm organic wastes to supply crop nutrients and biological control of pests and diseases. Farmers in developed regions were the first to devise the legislation required for organic production, but many farmers in developing regions, where agriculture is often less intensive anyway, are now seeing an opportunity to increase their profit margin at little extra cost. In Europe, the land devoted to organic farming practices has grown rapidly since the mid-1980s, reaching nearly 10% in Sweden and Austria. The regulations for organic cattle farming are considered by some to be extreme; for example, modern farming practices of zero-grazing cattle and embryo transfer are forbidden, as are the more contentious issues such as genetic engineering. However, in the absence of an accurate knowledge of the precise risks of many modern farming practices it can be said to provide the best possible assurance to the consumer that the production of the food that they purchase has not harmed the environment or the animals and will not harm themselves.

Conclusions

Since their domestication 8000–10,000 years ago, cattle keeping has been diversified to different parts of the globe. Cattle were an easy way of using land inhabited by native peoples and animals for the production of meat, milk and other goods needed by settlers during the periods of colonization in the last millennium. Now that their experience has spread to nearly all parts of the globe, it is necessary to examine the relationship between humans and cattle and decide whether it is the best way to feed the population, whilst at the same time maintaining a high quality environment and regional culture. Some systems of cattle production that have been developed are ecologically unsustainable and lead to deterioration of the environment. Others offend certain peoples' moral or religious beliefs, but many of today's systems make an important contribution to the nutrition of the human population by using land in a sustainable and worthwhile manner. The future will bring greater control of cattle production, preserving those systems that benefit society and outlawing those that have detrimental effects on the region in which they are practised.

Further Reading

Clutton-Brock, J. (1999) *A Natural History of Domesticated Animals*, 2nd edn. Cambridge University Press, Cambridge and New York.

Felius, M. (1985) *Genus Bos: Cattle Breeds of the World*. MSD Agvet, Rahway, New Jersey.

Rifkin, J. (1994) *Beyond Beef: The Rise and Fall of the Cattle Culture*. Thorsons, London.

Feeding Methods

2

Feeding the Calf

Colostrum consumption

Dairy calves are usually weaned from their mothers at 12–24 h postpartum, by which time they should have consumed colostrum from their mother. Colostrum is a mixture of blood plasma and milk, which is particularly valuable to newborn calves for its high concentration of immunoglobulins (Table 2.1). It also contains more vitamins than milk, the extent of which depends on the vitamin status of the cow. The permeability of the calf's small intestine to immunoglobulins declines rapidly after about 12 h, and is very low after the meconium has been passed (Roy, 1990). This is a black–green sticky substance that has accumulated in the gastrointestinal tract during the calf's life *in utero* and is usually passed at about 24 h after birth. Licking of the calf's anus by the cow during suckling probably stimulates its expulsion. The calf does not normally produce endogenous immunoglobulins for about 2 weeks, or perhaps 4 or 5 days earlier if they are deprived of the so-called 'passive

Table 2.1. Composition of colostrum and milk.

	Colostrum	Milk
Fat (g kg^{-1})	36	35
Protein (g kg^{-1})	140	33
Immunoglobulin (g kg^{-1})	60	1
Casein (g kg^{-1})	52	26
Albumin (g kg^{-1})	15	5
Lactose (g kg^{-1})	30	46
Vitamin A (μg g^{-1} fat)	42–48	8
Vitamin B$_{12}$ (μg kg^{-1})	10–50	5
Vitamin D (ng g^{-1} fat)	23–45	15
Vitamin E (μg g^{-1} fat)	100–150	20

immunity' transferred in the colostrum. It is not until 8 weeks after birth that the calf serum γ globulin levels are normal. It is therefore essential that the cow's colostrum be consumed within the first day of life, preferably about 7 kg for each calf, which is sufficient to provide about 400 g of immunoglobulin. European Union (EU) regulations specify that calves should receive colostrum within the first 6 h of life. As the immunoglobulin content of the cow's milk declines quite rapidly over the first few days of lactation, the cow should not be milked before parturition. The adequacy of immunoglobulin intake can be tested from a blood sample to which zinc sulphate is added, with the degree of turbidity in the sample indicating the extent of immunoglobulin absorption by the calf from the colostrum.

Inadequate colostrum consumption can arise either from lethargy in a cow or her calf as a result of a difficult calving, or from the calf having difficulty in locating the teats. In the latter case the herdsperson should assist, guiding the calf to the teat whilst expressing colostrum from the teat to encourage the calf. In extreme cases, or where the calf's mother cannot provide colostrum, and no colostrum is available from other newly calved cows, an artificial colostrum can be made from a raw egg added to milk, with some castor oil to act as a laxative. This will not contain relevant immunoglobulins for the calf, but the egg albumin is rapidly absorbed and some protection against septicaemia is provided. After removal from their mothers, the calves should continue to be fed colostrum, which is produced by the cow until the fourth day of lactation.

Artificial milk replacers

From approximately 4 days of age, many calves are fed 'artificial' milk – milk replacer – for about 5–6 weeks. Milk replacers are based on dried milk powder, produced from surplus cows' milk. Cows are now capable of producing many times the requirements of one calf, but feeding dried milk is not quite so circuitous a practice as it might seem at first sight, since economic production is difficult to achieve if the calf is allowed to suckle some of the cow's milk, with the rest being taken for human consumption. There seems little doubt that the welfare of both cow and calf would be at least temporarily improved by suckling, but even if we could restrict the calf's suckling and take the surplus for human consumption, the suckling and bunting of the udder by the calf elongates both the udder and the teats, making it difficult for the cow to be milked by machine. Also if the calf suckles for several weeks the eventual separation is more stressful for both parties. The separation of the cow and calf appears to be less stressful for the Holstein–Friesian cow than other breeds. In some calf-feeding systems, milk is taken from the cow by machine to be fed to the calves in buckets, particularly if milk powder is expensive. This system avoids adverse effects of calf suckling on the cow and in particular the stress of the eventual separation, while retaining the benefits of providing fresh milk for the calf.

Digestion in the young calf

The liquid diet that the calf consumes bypasses the rumen because of an oesophageal groove, which closes when the calf consumes liquids, and enters the abomasum. Here it forms a curd, with the whey passing through into the duodenum. The digestion of protein by the neonatal calf is restricted to casein, for which the enzyme rennin is produced in the abomasum. If the milk protein is denatured, such as in pasteurized or UHT milk, it takes a long time for the clot to form and the calf may get diarrhoea or scours, often with an accompanying *Escherichia coli* infection. In the older calf, when the parietal cells of the abomasum start producing hydrochloric acid, pepsin assists in protein degradation. Milk fat from the curd is digested by salivary esterase. Calves can digest a wide variety of fats, so the addition of vegetable-based fats to skimmed milk powder is possible. New spray-drying technology has allowed fats to be incorporated more effectively in an emulsified form into skimmed milk powder. The digestibility of fats of vegetable origin is still about 5% less than milk fat, believed to be caused by the binding of bile acids by insoluble calcium phosphate in the small intestine.

As milk surpluses in Europe have been brought under control there has been a need to find the best method of replacing at least some of the protein in the milk powder with vegetable-based products, such as soya protein. This has to be extracted with alcohol during heating to remove a specific antigen that can cause severe allergic reactions. The protein in soya is less digestible than milk protein since the main abomasal protein-degrading enzyme, rennin, is specific to casein, which is the main protein in milk. Milk protein is usually 85–95% digestible, whereas if vegetable-based replacers are included at 30–50%, the protein digestibility may be only 65–75%. This may still give acceptable calf weight gains, but reduced growth in the early period of the animal's life may take a long time to be regained. Calves fed soya protein concentrate are more likely to scour or become bloated. Whey protein concentrate is an alternative to vegetable-based replacers but the protein digestibility is still less than for skimmed milk.

Feeding milk powders

A conventional milk powder for calves contains at least 60% dried skimmed milk, 15–20% added fat and added minerals and vitamins. Skimmed milk is a by-product of butter manufacture and is composed mainly of lactose and milk proteins, principally casein. Whey, which is a by-product of cheese making, is composed of lactose and the whey proteins, principally albumin and globulin.

Conventional skimmed milk powder provides all the nutrient requirements of the newborn calf. Only some antibacterial agents in milk, such as the lactoperoxidase complex, are denatured by the processing, but these can be added artificially. Conventional skimmed milk powders have to either be fed to the calves in a bucket, which may have a teat attached to it, or by a machine which mixes it up 'on demand'. Its keeping life is short, compared to acidified milk powders, and it is normally fed once or twice a day in buckets. If large quantities of milk powder are fed, or the calves are small, then feeding just once a day is more likely to lead to milk spilling into the rumen and causing diarrhoea (calf scours). In the first week of the calf's life the curd only persists for about 8 h after feeding. Once-a-day feeding also results in less gastric and pancreatic secretions than twice-a-day feeding. Calves normally suckle about five times a day, so once-a-day feeding is likely to overload the digestive tract, with its only advantage being a reduced labour requirement.

The amount of milk powder fed to a calf will determine its growth rate, and the stockperson must decide whether a faster growth rate at an earlier age justifies the high cost of additional milk replacer. The milk replacer only provides about 0.5% of the total metabolizable energy (ME) requirements of the dairy heifer from birth to first calving, but a high rate of growth initially will lead to strong, healthy calves. Calves require about 360 g milk replacer daily in two feeds of 1.5 l, after finishing their colostrum feeding at day 3 of age, for them to grow at a rate that will allow them to calve at 2 years of age. Later this can be increased to up to 1 kg milk replacer daily, diluted to 7 l with water, which should be sufficient for 50 kg calves to grow at about 0.7 kg day^{-1}. Alternatively one feed of a slightly stronger mix can be provided.

Bucket feeding

Milk replacer is reconstituted once or twice daily in bucket feeding systems. It is normally mixed with one half of the water at 46°C and then the remaining water added is either hot or cold, as required to achieve a final temperature of 42°C. This temperature is important to achieve optimum digestibility.

Feeding from buckets is not a natural process for a young calf, whose instincts direct it to suckle from birth. However, calves can be successfully trained to drink from buckets using two upturned fingers immersed in the milk, to simulate the mother's teat. When the calf has learnt to suck milk around the fingers, they can be withdrawn gradually and the calf will drink unaided. This process is sometimes difficult for the calf to learn, as suckling from a teat involves *squeezing* it with the tongue to express the milk, starting with pressure applied at the base of the teat and with the tongue transferring the pressure to the tip of the teat. Some sucking is involved as well, but the main milk extraction mechanism is by squeezing the teat. Calves are often reluctant to put their heads into a bucket, since they naturally suckle with their head in the horizontal position (which aids oesophageal groove closure; Phillips, 1993).

The absence of teat suckling encourages calves to suckle other suitable protruding objects, especially immediately after they have been fed their milk in buckets. They will 'suckle' the bucket handles, any bars or other parts of their pen and, if they are housed so that they have access to other calves, their neighbours' tongues, ears, tails and navels. Navel sucking keeps the navel moist and encourages bacterial invasion, often leading to septicaemia (navel ill/joint ill). The sucking behaviour often persists when the calves are older in the form of urine or prepuce sucking and tongue rolling in males, and intersucking of teats in females (Rutgers and Grommers, 1988). The stress caused by the absence of a suckling stimulus is exacerbated by individual penning of the calves. Individually penned calves may also spend a lot of time licking objects in the pen, which may be caused by inadequate mineral supply, principally sodium (Phillips *et al.*, 1999). Sodium requirements are increased by the stress caused by individual penning and additional salt can be added to the concentrate, but this will increase water consumption and urination, so that it is important to regularly ensure that the calf has enough clean bedding.

In summary, the bucket feeding system offers the potential to control the calf's intake and minimize cross-infection between calves by limiting contact between individuals. However, the absence of suckling encourages the calf to suck objects in its pen, which can lead to abnormal behaviour in later life.

Alternative feeding systems

Increasingly alternatives to bucket feeding are used to avoid the obvious distress to the calf as a result of the lack of a suckling stimulus. The simplest is to raise the height of the bucket and put a teat on the bottom. Alternatively a teat may be suspended in the milk, but this does not completely satisfy the suckling drive. Stimulating sucking behaviour encourages the production of salivary esterase, so the stereotyped oral behaviours of bucket-fed calves, which are often described as purposeless, may actually be of nutritional benefit to the calf.

Group-feeding systems

Milk may also be fed to calves housed in groups by teats attached by a length of pipe to a large container, provided that the milk is acidified to keep it from spoiling. Usually weak organic acids are added to reduce the pH of the milk to about 5.7, since casein will clot if the pH is reduced any further. Stronger acids can only be added if the powder is whey based, to give a pH of about 4.2 (Webster, 1984). Good calf growth rates are achieved, even though the milk replacer does not clot in the abomasum. Acidified milk replacers are usually made available *ad libitum,* and intakes may be 20–30% more than if the calves are fed from buckets. The intake depends on the temperature of the milk, and whereas the milk is warmed to 42°C for bucket feeding, it is available at ambient temperature for group feeding of acidified milk replacer *ad libitum.* If the milk is at 10–15°C or less, it chills the calf, intakes decline and so does the calf growth rate. Not all calves respond equally, and the stockperson needs to keep a careful watch for any calves that are not thriving on the system. The main advantage of group-rearing calves and offering them acidified milk replacer is that it saves labour, but some of the time saved must be used to check the calves regularly.

The keeping life of acidified milk replacers is approximately 3 days, depending on the temperature, and containers and pipes must always be cleaned in between feeds. Calves reared in groups spend less time sucking objects in their pen, but there is more opportunity for cross-infection, particularly via the navel (Dellmeier *et al.*, 1985). There should be no more than ten calves in each group and teats should be inspected regularly for damage. The calves play with the teat a lot, particularly when milk powder stops being fed, and the teats should preferably be removed at this time. During milk feeding the teats should be securely fastened at a height of about 600 mm from the floor and the tube leading to the milk store should have a one-way valve. There are claims that calves on acidified milk replacer are less likely to scour, despite their high intakes, probably because the calves consume their milk in small meals, so there is less likelihood that milk will enter the rumen, and also the acidification reduces the likelihood of bacterial contamination of the milk (Roy, 1990). Nevertheless, the feeding system should be cleaned out regularly with hot water and a mild disinfectant.

In summary, calves fed on acidified milk replacer offered *ad libitum* will consume more milk powder and less solid food than calves fed on restricted milk powder fed in buckets (Table 2.2). They therefore grow more rapidly, but the cost of that growth is more. As solid feed costs less than one half that of milk replacer per unit of energy, the producer must decide whether the benefit of extra growth is worth paying the extra feed costs. Pedigree breeders may well decide that it is, but those producing less valuable animals are likely to minimize input costs.

Another alternative to restricted milk replacer or acidified *ad libitum* milk replacer is to use machines that make up milk powder for each calf 'on demand'. Calves in groups can be fed a pre-programmed amount of milk replacer at body temperature. The milk replacer has to be carefully formulated so that it flows freely in the machine's hopper, and safety arrangements considered so that power or water failures do not leave the calves without

Table 2.2. A comparison of the feed intake and growth of calves on restricted milk replacer fed in buckets and acidified milk replacer fed *ad libitum*, from birth to 3 months of age.

	Restricted milk replacer (buckets)	*Ad libitum* acidified milk replacer (teats)
Milk substitute intake (kg)	11–16	20–30
Concentrate intake (kg)	115–130	105–120
Hay intake (kg)	5–9	6–10
Live weight gain (kg day^{-1})	0.5	0.8
Relative cost per kg gain (restricted system = 100)	100	125

their feed supply. The machine recognizes each calf by an electronic key suspended around the calf's neck. The system is expensive, but has the advantage that all calves will consume the programmed amount of milk, which is delivered via a teat so that the calves do not get frustrated by the absence of a suckling stimulus.

Colostrum can continue to be fed to calves after the period of suckling, as cows produce about 50 l of colostrum in the first 4 days of lactation, which is considerably more than a single calf can consume in that time. It will naturally ferment if stored in barrels, and the resulting soured product has a pH of about 4.5 and will keep for about 3 weeks. It supports good calf growth rates, but the acceptability of the product is variable and it is better fed to calves at times of the year when ambient temperatures are not too high. Calves on soured colostrum are unlikely to get diarrhoea, and in particular there are fewer rota and corona viruses contracted.

Introducing solid feed

For the first few weeks of a calf's life the milk replacer feeding systems described above will provide sufficient nutrients for near maximum growth rates in the calves. However, because of the high cost of milk replacer and the requirement by the calf for some solid food, it is usual to offer concentrate feed from about 1 week of age. Legally in the EU all calves over 2 weeks old must have access to fibrous food (at least 100 g at 2 weeks of age to 250 g at 20 weeks). This allows the rumen papillae and rumen motility to develop normally. Many calves that do not have access to solid food develop hairballs in their rumen, caused by the consumption of hair during their licking and grooming activities and the lack of rumen motility to transfer them through to the abomasum.

Concentrate food offered can be either a small pelletted (3–5 mm) compound food or a 'coarse mix', which is a blended mixture of concentrated ingredients without any attempt to bind them together. Both foods produce similar calf growth rates, but the pelletted products have the advantage that selection by the calf is not possible and there is little feed wastage, providing that the pellet is sufficiently hard. If it is too hard the calves eat less as they find it hard to chew. Coarse mixes tend to be dusty, leaving a mixture of dust and saliva on the bottom of the bucket. Farmers may have difficulty in obtaining or storing the correct ingredients for a coarse mix, and the price paid will usually be higher than that paid by the feed compounder. With both feed pellets and coarse mixes, but particularly the latter, it is best to offer only a little bit more than the calves are eating daily, but careful management is necessary to avoid restricting some calves. If milk is offered *ad libitum* only a small amount of concentrate feed will be consumed during the first few weeks, and a greater check to growth rate occurs at weaning. However, it is important for the calves to get used to having concentrate available to them. Intake increases over the first few weeks of life until it is normally about 1–1.5 kg at 5–6 weeks of age.

After this farmers may restrict intake of dairy heifer replacements to 2 kg at this stage, but bull calves may continue to receive concentrates *ad libitum* to achieve maximum growth rates.

The crude protein content of the concentrate feed is important, and should be at least 18% because calf growth rates at this time are high relative to their intake. The proportion of crude protein that is degraded in the rumen should not be more than about two-thirds, with feeds such as soya being suitable for this purpose. Mineral supplements should be added to prevent calves searching for minerals, particularly if they are kept in individual pens. In particular, sodium, potassium, magnesium and chlorine may become deficient when a calf scours, and extra calcium and phosphorus are important for bone growth. The calcium source is important in determining availability, with the calcium in milk being particularly well absorbed (*c*. 95%). In this case maintaining an adequate mineral status depends to a considerable extent on the concentration of minerals in the milk replacer. Iron is the trace element most likely to be deficient, with two iron-binding proteins, lactoferrin and transferrin, being responsible for the transmission of iron from cow to calf and the restriction of iron availability to any bacteria that might enter the milk. Inadequate concentrations of lactoferrin and transferrin in milk are common, and responses to iron supplements can be observed as early as 2 weeks of age. Later, when the intake of solid food has increased, blood haemoglobin levels are restored and supplementary iron is unnecessary.

Forage for young calves

The offering of forage to young calves is controversial, since too much reliance on low-quality forage reduces growth rates. Calves that rely too much on coarse forage develop a large rumen at an early age (Pounden and Hibbs, 1948), which makes them look 'pot-bellied', and this can result in low weight gains (Thomas and Hinks, 1982). However, forage is important for the creation of stable rumen conditions and in particular a high and relatively constant rumen pH. The addition of 15% straw to an otherwise all concentrate diet for calves will increase intake, demonstrating that calves eating solid food have a requirement for fibre in their diet. The encouragement of stable rumen conditions at an early age will help to avoid any nutritional disturbances, such as bloat or scouring. EU regulations now require that all calves over 2 weeks have access to fibrous food (at least 100 g day^{-1} per calf must be available at 2 weeks, increasing to 250 g day^{-1} per calf at 20 weeks).

Some farmers use concentrate pellets that contain chopped straw, but it will not have the physical properties of long forage that are important for rumen development. For this reason the cereal ingredients in concentrates should be rolled or flaked but not ground, as this also does not stimulate rumen activity.

Forage offered to calves should be of the highest quality and palatability. The best is fresh, leafy grass, but this is difficult to provide in sufficient quantities for large numbers of calves. In some countries calves are let out to pasture during the day and brought in at night to be fed milk replacer. This provides the calves with exercise, stimulation, companionship and nutritious food, but it uses more labour than conventional systems. Allowing calves access to fresh air greatly reduces the risk of disease transmission, but it would be unwise to let them outside in inclement weather unless shelter is available.

Fresh grass is eaten avidly by calves from an early age, and when fed to calves in individual pens it reduces any behavioural problems caused by insufficient suckling activity. Indeed the latter is rarely a problem once an adequate intake of solid feed has been achieved, so forage must be palatable to encourage the calves to start eating it early. This does not necessarily mean that it should be of high digestibility, since rumen capacity does not limit intake in the first few weeks of life (Kellaway *et al.*, 1973), but it should be dust-free, sweet smelling and not too dry. Silage tends to be avoided by very young calves, but good quality hay or straw will be eaten in sufficient quantities. Small bales of hay are easier to handle than silage in the calf house and it keeps fresh for longer, but many cattle farms have changed from haymaking to silage in the last 30 years. The crude protein content of hay is usually less than silage, but the higher intakes of hay may offset this to give an adequate crude protein intake from forage. Straw is suitable for calves, provided that it is not dusty. It rarely has any offensive odours that would reduce its acceptability to calves. The palatability of straw can be improved by spraying it with molasses, but producers should beware of the high potassium content of molasses that may reduce sodium availability, leading to licking problems referred to above. Sugarbeet pulp is a good source of roughage for calves, as it is highly digestible and supports a high rumen pH.

Weaning calves

The age at weaning may determine the ability of the calf to grow adequately in its early life. Weaning at too early an age and without adequate, high-quality solid food may lead to reduced growth rates for much of the calf's first year of life, although there is no indication that this could affect carcass composition in beef cattle (Berge, 1991). There are no wild cattle to give us an indication of the natural weaning age of calves. In feral cattle, where there is minimal interference by man, female calves do not naturally sever the bonds with their mother even when they are mature (Reinhardt and Reinhardt, 1981). Male calves will leave the matriarchal group soon after sexual maturity. Suckling can continue until the cow prepares to have her next calf, which may be one or more years after the birth of the first calf. In domesticated cattle weaning strictly speaking refers to when calves are removed from their mothers, usually at about 1 day of age, but most people consider weaning as the time when calves cease to receive milk or a milk substitute. The high cost of milk substitutes and whole milk for feeding to calves, as opposed to sale for human consumption, encourages farmers to transfer calves to solid feeds at an early age. In addition calf rearing is time consuming, especially if they are fed once or twice per day. Although much concern is expressed for the welfare of calves removed from their mothers at such an early age, group rearing on solid food is probably better for the calves' welfare than rearing on a milk-based diet in individual pens.

Economics normally dictate that calves stop getting milk replacer at between 5 and 8 weeks of age. Small calves, in particular those from Channel Island breeds, should be weaned at the later age otherwise they could have a high mortality rate. It is easy to overfeed small calves with milk, so they need to be fed proportionately less and weaned later. Calf mortality should not be more than 6% for homebred calves. For calves that are sold through an auction market, the mortality rate is typically about 14%. The risk of mortality is particularly high in the spring because of the accumulation of pathogens, especially *E. coli*, in the calf house.

When deciding the optimum age at which to wean calves, most farmers take into consideration the size of each animal and how much concentrate feed it is eating. Often calves are weaned in batches, for convenience, so for some calves it may be too early and some too late. If a farmer can wean according to weight, Friesian calves should be at least 50 and preferably 60 kg before milk is withdrawn. The calves should be eating a daily minimum of 0.6 kg and preferably 1.0 kg concentrate. Because they are usually weaned in batches to form their subsequent groups, there will be a range of ages and weights. There is no definitive scientific evidence that gradual weaning over 1 week is better than abrupt weaning, but, as a general rule, sudden changes to the diet should be avoided. It takes some time for the digestion system to adapt after changes to the diet have been made and sudden changes may stress the calf. Gradual weaning may be achieved by weakening the mixture or omitting one feed per day. The latter can lead to restless calves at the time when the second feed used to be offered, and it is probably better to gradually reduce the amount that the calf is fed. Calves fed acidified milk replacer *ad libitum* can have the pipe to the milk store disconnected for a period of the day, but this can lead to considerable disturbance every time it is reconnected.

Restricted suckling systems

In developing countries cows often suckle their calves as well as being milked, which is known as restricted suckling (SandovalCastro *et al.*, 1999). There are several reasons why farmers adopt this practice.

1. Milk powder is expensive.
2. Cow productivity is increased.
3. Zebu (*Bos indicus*) cows do not readily let down their milk unless their calf is present, so restricted suckling is essential if high yields are to be achieved. With such cows the calves must suckle the cows first, and any surplus can be milked by machine for sale after the calves have been removed.

A common problem with restricted suckling systems in the tropics is that the calves are removed before they have had sufficient milk, which maximizes the saleable milk yield but often leads to high calf mortality.

Bos taurus cattle will let down their milk in the absence of their calf, so the calves can suckle the cows after the milk for sale has been extracted by machine. In this case the calves will be drinking high-fat milk, because during milking the fat globules are retained in the alveolar tissue until the latter stages of milking. The milk extracted for sale under such circumstances is of reduced fat content. The total milk production by the cow is increased, partly because the calf is able, by bunting and pushing the udder, to extract residual milk that would otherwise be left by the machine. However, the increase in milk yield persists even if the calf stops suckling, suggesting that the long-term production potential has been increased by a period of suckling. The extraction of residual milk helps to reduce mastitis, as there is only a small milk reservoir permanently in the gland to support bacterial growth. The amount of milk obtained by the calf can be regulated by controlling the number of suckling periods each day, normally one or two, and their length, which is usually about 20 min. It is nevertheless difficult to ensure that calves have adequate milk and the optimum quantity is removed by machine for sale.

A particular problem with restricted suckling systems is the considerable extension of the postpartum interval to rebreeding. The extent of this depends on the amount of suckling – *ad libitum* suckling may cause complete lactational anoestrus in some cows and will on average extend the interval to first oestrus by 34 days, compared with systems where cows are milked only by machine. Twice-a-day suckling still results in a significant lengthening of the interval to first oestrus postpartum, but there is much less effect if the calves are only suckled once a day. The mechanism by which the phenomenon occurs is a reduced luteinizing hormone output by the pituitary gland. The effect requires suckling to occur, the presence of a muzzled calf having a much-reduced impact. Attempts to overcome the suckling inhibition of oestrus by temporary weaning or the use of exogenous hormones have not been particularly successful. This is in part because some of the inhibitory effects of suckling on cow reproduction are caused by a long-term reduction in cow body condition. The extra milk production required for the cow to satisfy both calf and machine is often not compensated for by providing additional feed, with the result that body condition declines considerably in early lactation.

Single-suckling systems

Calves usually grow rapidly when they remain suckling their mother, providing that she has enough grass to support a reasonable milk yield. Some of the traditional beef breeds may produce less than 1000 l per lactation, which is only just enough for a single calf. The single-suckling system is best suited to extreme hill conditions, since in better conditions the cow should be able to produce enough milk for twins. In the hill farms, spring-calving cows can be fed low-quality rations during the winter, based on straw and a small daily concentrate feed, and will then restore their body weight when they are turned out to pasture in spring with their calves. Calving is easier if the cows have been fed on a low-energy winter ration. The calf performance will be quite adequate provided sufficient grass is available, so spring calving is favoured in the more extreme hill conditions. Autumn-calving cows need more feed in winter if they are to sustain milk production for their calves, although if they lose weight over the winter, they will compensate at pasture. If feed availability is restricted in winter, autumn-calving cows are less likely to conceive, whereas spring-calving cows are on a rising plane of nutrition when mated in summer and more readily conceive (Wright *et al.*, 1992).

Feeding Growing Cattle

In Britain, cattle are usually fed conserved feeds for approximately their first 6 months of life, as they are vulnerable to inclement weather until this time. If the opportunity arises, offering them fresh cut grass indoors enables them to be turned out earlier as they will start grazing sooner.

After weaning, many farmers prefer to feed silage to their calves until they are turned out to pasture, as it is easier to handle, more readily available and of greater nutritional value than hay or straw. Intake should steadily increase as the calf gets older and develops the rumen microflora to digest the silage, until the calf is eating about $12–15$ g kg^{-1} live weight. The calves respond particularly well to supplementary protein that is protected from rumen degradation until they are about 5 months of age, because of their rapid growth rate at this time. Fishmeal is a good source, although it is not always readily accepted. The requirement for rumen-undegraded protein depends on forage quality – if the forage is rapidly broken down by the rumen microbes then the ready availability of volatile fatty acids in the rumen to supply energy to the calf will need to be complemented by such protein. Usually silage will supply sufficient rumen-degradable protein (RDP) for maximum microbial growth, as the silage protein is about 80% rumen-degradable. Formaldehyde-treated silages may be less degradable, perhaps only 65%.

For high growth rates in older calves on a silage-based ration, supplementary concentrates should be continued at about 1 kg day^{-1} if the silage is highly digestible (68–70 D-value), and 2 kg day^{-1} if the silage is less digestible (60–68 D-value). They should include a protein source, such as soybean meal, and may include a source of protein that is of low rumen degradability, such as fishmeal, if the forage is high energy. The more concentrates that are fed, up to perhaps 5–6 kg day^{-1} per calf, the faster the calves will grow, but they will then grow slower after they have been turned out to pasture. The feed additive sodium monensin will improve calf growth rate by increasing the proportion of the rumen volatile fatty acid propionate at the expense of acetate production. It can now be administered as a slow-releasing bolus at pasture, which avoids the check to growth that used to occur if it was removed from the calves' diet when they were turned out to pasture.

When calves are turned out, it is important to continue offering concentrate feed for 2 or 3 weeks to minimize any sudden change in the type of feed available for the rumen microorganisms. The concentrate feed need not have a high protein concentration, as a result of the high protein content of grass, and could be based on cereals, rolled to improve digestibility and with a vitamin and mineral supplement added. This supplement should be sufficient to supply 1000,000 IU vitamin A and 2000,000 IU vitamin D per tonne. The concentrate feed should be particularly fed during wet weather as intake is considerably reduced when the grass is both young and wet. It is also useful to offer hay in a rack to buffer variation in the intake of fresh grass. Hay will keep for several weeks compared with a few days for silage, and will help to maintain a high rumen pH for adequate fibre digestion. If the grass is young and leafy young calves should be only allowed out for a few hours each day for the first week, otherwise they will get diarrhoea caused by the limited amount of fibre in their diet. They may then lose condition as the limited time that the grass remains in the gastrointestinal tract reduces the nutrient absorption. Also the gut motility declines which may reduce intake.

For autumn-born calves turned out in April, the stocking rate on good quality pasture should be up to 15 calves ha^{-1}, as individual consumption will be low. If the sward is of low quality and little fertilizer is applied, the rate has to be reduced. It is most important to keep the herbage in a young, leafy state otherwise its nutritional value declines rapidly. This high stocking rate will not normally need to be relaxed until June, when it could be halved by the introduction of silage aftermaths. This will help to reduce the risk of parasite infection, and if possible the calves should be moved to pasture that has not been grazed that year. A final reduction in stocking rate could be made in August to about 5 animals ha^{-1}, perhaps after a second cut of silage or when in-calf heifers are removed to enter the dairy herd.

A salt lick should be provided if the cattle are to achieve high growth rates at pasture, particularly in areas where the pasture is naturally deficient in sodium or high temperatures make the cattle lose sodium by sweating. The salt lick should be placed in a container so that it does not get contaminated with earth, and preferably should be protected from rain. Clean water is also essential and water from a stream or pond may be diseased, so mains water should be provided. It is unwise to rely on cattle gaining their water supply from natural sources, as they will damage the banks and pollute the water with excreta. If cattle are detected with infections, such as tuberculosis, on a farm, it would be prudent to clean out all water troughs immediately and thoroughly. Cattle drink mainly from the surface of the troughs, and in large capacity troughs, the turnover of water at the bottom of the trough may be negligible. Organisms such as *Mycobacterium bovis*, which causes tuberculosis, can survive for up to 1 year in water. If there is no obvious water source for badgers on a farm, or if there are signs of badgers accessing water troughs (such as scratch marks on the sides of the trough), measures should be taken to prevent badgers from using troughs by raising them to a height of at least 0.8 m and using troughs that make it difficult for them to climb up the sides.

Spring-born calves, which are usually turned out to pasture when younger than autumn-born calves must be stocked at a higher density, up to 30 calves ha^{-1}, and some shelter should be provided in exposed areas. There is a limit to the extent to which grazing can be intensified in an attempt to keep the pasture in short, leafy condition, as the calves are quite selective grazers. A leader–follower system is also suitable for growing cattle, with about 8–10 paddocks that will allow for a minimum period of 21 days regrowth before each grazing. The system is described in more detail in Chapter 4.

In autumn, the grass is usually of poor quality, particularly in arid regions. Even if the rainfall is evenly distributed over the year the increased grass transpiration rate in summer will deplete soil water reserves and reduce grass growth. A concentrate supplement of rolled cereals will help to maintain growth rates during excessively dry or wet periods. A mineral supplement should be provided too. Forage supplements can also be offered during the summer if dry weather reduces the amount of herbage on offer. However, if there is adequate grass available, supplementary concentrates will only substitute for grass intake and the profitability of the cattle rearing system will decline.

The cattle should be housed whenever the conditions dictate, i.e. when the grass is too short (less than about 8 cm tall), or when the land is too wet and continued grazing would damage the sward and only support low herbage intakes. Young cattle do less damage to a sward in wet conditions than adult cows, because of their light weight, but the temptation to leave them out late into the autumn should be resisted if the weather is inclement.

When cattle are rehoused, it is preferable to feed them silage, rather than hay, as it is of better quality in terms of its energy and protein content. With reasonable forage quality, no more than about 2 kg of concentrates needs to be fed daily to dairy heifers to achieve adequate growth rates for them to calve at 2 years of age. If the heifers are required to calve the following autumn, they will have to be inseminated in the mid-winter period. After this time the concentrate feeding can be reduced in expectation of compensatory growth during the following summer grazing. However, the heifers must achieve an adequate pre-calving weight (500 kg for a traditional small Friesian, up to 630 kg for a modern, large-framed Holstein–Friesian), otherwise first lactation yields will be reduced as the animal has to divert too many nutrients to continued growth. In addition a heifer does not settle well into a herd if she is much smaller than the older cows. However, the heifer must not be too fat or the incidence of dystokia will be increased, and a herd manager should aim for a body condition score of 3–3.5 at calving.

For a Friesian heifer, the insemination, either natural or artificial, should be when the animal has reached a weight of at least 330 kg. If less than this the heifer will still conceive, but will have difficulty reaching the required pre-calving weight.

For a steer that is expected to be marketed at the end of its second winter at 18 months of age, more concentrates will have to be fed during the latter part of the winter than to a dairy heifer. Inadequate concentrate feeding leads to slower growth and a longer finishing period (Table 2.3), resulting in considerably increased total feed requirements but a heavier slaughter weight. The producer must decide whether the potential to finish at a heavier weight, in the example given 50 kg heavier, is worth the extra expenditure on feed and other variable costs. Often it is not, because the feed is inefficiently converted into live weight at this advanced stage of growth.

Many beef cattle rearing systems finish cattle at 18 months of age, with two winters indoors and a summer at pasture in between, but they often fail to produce a finished animal in this period. The main reason for low growth rates is inadequate growth in the summer grazing period. Problems include:

- not adopting a high enough stocking rate in the early part of the grazing season, so that the herbage becomes rank and of low quality;
- inadequate control measures for parasitic infection (ostertagiasis, parasitic bronchitis and fasciolosis);
- exceptionally dry or wet summers when the cattle have low intakes of grass.

In the winter the main problem is not adequately compensating for low-quality forage by offering additional concentrates (Table 2.4). Approximately 1 t of concentrates will need to be fed to each animal in total, including the concentrates that have to be fed before the second winter housing.

Table 2.3. The effect of amount of daily concentrate fed on the growth and total feed requirements of beef cattle.

	Rolled barley (kg DM day^{-1})		
	2.0	2.4	2.8
Silage intake (kg DM day^{-1})	5.3	5.1	4.8
Live weight gain (kg day ^{1})	0.7	0.8	0.9
Finishing period (days)	285	220	165
Slaughter weight (kg)	525	500	475
Total barley requirement (kg)	670	620	525
Total silage requirement (t)	6.0	4.4	3.2

Table 2.4. The effects of silage quality on feed requirements to achieve a growth rate of 800 g day^{-1} in steers.

	Silage 'D' value		
	55	60	65
Rolled barley (t per animal)	0.84	0.59	0.26
Silage (t per animal)	3.4	4.4	5.5

Often the quality of the ration is not sufficient for the cattle to finish within 18 months, and they have to be turned out to pasture again for a second summer. This is particularly the case for animals of large breeds, such those from continental Europe. Although these have a faster growth rate than smaller breeds, they will still reach the same level of fat cover at a later date than small breeds of cattle.

If the steers are to be finished after a second summer at pasture, the feeding in the second winter can be much reduced to make good use of compensatory growth in the final summer. Failure to do this is a common reason for low profitability in cattle finished at 2 years of age. Suitable rations for an animal at 250–300 kg at the start of its second winter would be about 20 kg silage (at 20–25% dry matter [DM] content) and 0.5–1 kg rolled barley daily. If straw is fed, the cattle will probably eat between 3.5 and 4.0 kg daily and more concentrate feed will have to be offered if the animals' growth rate is to be maintained, perhaps 3 to 3.5 kg day^{-1}.

In their second summer the stocking rate of the cattle will need to be much less than the previous summer, on a well-fertilized sward perhaps 3–3.5 steers or heifers per hectare. A leader–follower system of grazing is well-suited to the requirements of cattle in a 24-month system, but requires at least small fields or preferably paddocks to allow the two groups of cattle (first year and second year) to be rotated around the pasture area. If a mixture of breeds is being finished, the removal of some of the smaller cattle breeds which finish early, such as the British beef breeds, part way through the grazing season will allow the stocking rate to be relaxed to provide more grass for the larger, later-maturing breeds, such as the Continental breeds.

In a dairy heifer replacement rearing programme, the pregnant heifers should be fed concentrate towards the end of the summer to encourage the rumen to develop a suitable microflora to digest the post-calving diet. Supplementary magnesium may be needed to avoid hypomagnesaemia in early lactation.

Beef cattle production from conserved feeds (storage feeding)

A logical solution to the problems that producers experience in obtaining adequate growth rates of beef cattle at pasture is to feed them conserved feed throughout their life. Advances in silage-making technology, a good market for the finished product and the importation of larger (Continental) cattle breeds with high growth potential resulted in the system gaining in popularity in Britain in the 1980s. In the 1990s the competition from outside Europe and a depressed beef market in the wake of the BSE (bovine spongiform encephalopathy) crisis led to lower cost systems being favoured. Nevertheless, on a worldwide scale, many cattle are fed on conserved feeds, at least during the finishing stage.

The system is most suited to exploiting the growth rate potential of bulls, which would often have to be castrated before they could be safely kept outside. The bulls are usually fed hay until about 12 weeks of age, after which they receive silage as their main forage. Silage feeding can be a problem in summer, since silage left for more than 1 or 2 days exposed to air has a high risk of secondary fermentation. This will be less of a problem if the farm has many animals and several long, narrow silage clamps, from which the silage is removed in neatly cut blocks rather than being pulled out by a tractor with a bucket or fore-end loader, so that the exposed face is minimized. Grass silage is a popular choice, although maize silage is used very successfully in the south of Britain and in many countries overseas. Maize silage, being of lower protein content compared to grass silage, needs to be fed with a high protein supplement, such as soybean meal. Alkali-treated straw can be used as an alternative forage, but this is no better than medium quality silage in terms of its ability to support cattle growth. If ammonia is used to treat the straw, the resulting product will contain much non-protein nitrogen, which is an adequate nitrogen source for the rumen bacteria. Adequate sulphur must be provided to allow the bacteria to construct sulphur-containing amino acids, such as methionine and cysteine. If straw alone is used, a non-protein nitrogen product, such as urea, can be poured onto the straw to increase its nitrogen content. Root crops, such as swedes, can be used to replace up to one-third of the forage, but if more than this is fed, the high-water and low-protein contents reduce the weight gain of the cattle (Fisher *et al.*, 1994). As with the finishing of beef cattle in winter on a silage/concentrate mix, if the forage is of poor quality it is usually better to increase the amount of concentrates fed, to ensure that the cattle finish at 12 months of age, rather than extend the period of finishing by another 2 months. The response to the additional concentrate is greater if the silage quality is low. Even a small difference in the amount of concentrate fed will have a marked impact on profitability (Table 2.5).

A cereal-based concentrate is adequate when grass silage is fed, but the cattle will respond well to the inclusion of high protein feeds that bypass the rumen in the first few months. However, the extra growth in the early period is partially offset by reduced growth later on (Table 2.6).

Table 2.5. The performance and profitability of storage-fed cattle offered two concentrate levels.

	High concentrate	Low concentrate
Concentrate intake (kg day^{-1})	2.3	1.9
Silage DM intake (kg day^{-1})	2.5	2.8
Live weight gain (kg day^{-1})	1.06	0.96
Slaughter weight (kg)	454	444
Finishing period (months)	12	14
Effective stocking rate (no. ha^{-1})	10.4	8.6
Relative gross margin (High concentrate = 100)	100	76

Table 2.6.   The weight gain of cattle offered different amounts of white fishmeal with their concentrate feed.

	2 kg barley	1.9 kg barley + 0.1 kg white fishmeal	1.8 kg barley + 0.2 kg white fishmeal
Live weight gain, 17–38 weeks (kg day^{-1})	1.1	1.2	1.3
Live weight gain, 38 weeks to slaughter (kg day $^{-1}$)	0.9	0.8	0.8

Beef cattle production from cereals

The efficient conversion of feed to live weight by cattle fed an all-cereal diet can sometimes be utilized to obtain high profits and a rapid turnover of cattle. In well-managed systems a food conversion ratio of only 5–6 kg concentrate is required for each kilogram of weight gain. The system was popular in the 1960s in Britain, when calves and cereals were both inexpensive (Preston and Willis, 1970). Up to one-sixth of cattle were produced in this way. As well as being fed an all-cereal diet, calves are usually weaned relatively early, at 5 weeks of age. They are then fed an all-concentrate diet, which for the next 7 weeks should have a crude protein content of 17% for maximum growth, including a source of protein that does not readily degrade in the rumen. At 12 weeks of age the crude protein content can be reduced to 14%, and then at 30 weeks it can be further reduced to 12%, which may be provided by rolled barley alone. Roughage is not usually provided, except that cattle may be bedded on straw and they will eat sufficient long fibre to avoid ruminal tympany (bloat) (Cheng *et al.*, 1998; Owens *et al.*, 1997). If low-fibre cereals are fed, such as maize grain, the inclusion of some fibre in the diet will increase weight gain, providing it is not more than approximately 20% of the dietary DM by weight. Less than 10% roughage in the dietary DM is likely to result in some cattle acquiring ruminal tympany. Ground maize, in particular, often leads to this condition, especially if the cattle have been away from the feed for some time, so it is better to feed whole cobs. Lucerne is not a suitable roughage as it induces ruminal tympany. Any changes to the diet should be made gradually, otherwise cattle may reduce their intake and then suffer digestive upsets when intake is restored.

Barley is the most common cereal used in cereal fattening systems for beef cattle. It should be dried to 14–16% moisture for optimum conversion efficiency and safe storage. Processing should aim to preserve the roughage content of the grain but break the seed coat so that the endosperm is exposed for digestion (Campling, 1991). Rolling or crimping is best, and grinding is inadvisable because of its destructive effects on the grains' roughage content. In the tropics the residual bagasse from the sugarcane plant, after the sugar has been extracted, is a suitable roughage source and sugarcane molasses may replace the concentrate successfully.

Cereal diets for bulls should have a mineral/vitamin supplement added to them, containing in particular vitamins A and D, salt and limestone, for maximum growth of the cattle.

Feeding the Dairy Cow

Introduction

The dairy cow has been selectively bred to produce considerably more milk than required by any calf, and if management is good, she will produce it for three-quarters of her life in the dairy herd. This requires a much greater intake of nutrients than the traditional high-fibre diet of cattle, in particular at the beginning of lactation, so nutritionists are constantly trying to devise means of increasing the cow's nutrient intake while maintaining stability in her metabolism.

Preparing for lactation

Dairy cows and heifers can be prepared for lactation by feeding some concentrates during the dry period. The amount of concentrates provided depends on the condition of the cow and the desired milk output, but would typically be 2–4 kg day^{-1}. In the case of a first-calving heifer, it can be achieved by adding a group of heifers to the milking herd, so that they can be fed concentrates when the cows go through the parlour to be milked. This makes them associate the visit to the parlour with something pleasant, and prepares them for milking, rather than them being suddenly faced with the procedure after calving.

Feeding a high-nutrient density diet during the dry period prepares the cow for milk production by firstly supporting growth of the rumen papillae, which takes about 5 weeks of exposure to cereals, and secondly, by allowing the cow to lay down additional body reserves that can be used in early lactation, when the nutrient requirements for milk production exceed those provided from feed intake. The preparation of rumen papillae is particularly important for first calving (primiparous) cows, which will consume less forage than older cows but may receive the same concentrate allocation in the parlour. The high concentrate:forage ratio makes these primiparous cows prone to metabolic disorders in the early part of their first lactation, particularly laminitis in the hoof. If they receive a complete diet they will get the same concentrate:forage ratio as older cows and will be less likely to develop metabolic upsets.

The disadvantage of high-concentrate feeding pre-calving is that cows that are fat at calving are more likely to develop dystokia and consume less feed postpartum than cows that are thin at calving. These rely more on catabolizing body tissue for their energy, and to some extent, protein requirements and their low intake makes them less likely to conceive. A large negative energy balance in the first 100 days of lactation increases the time to first ovulation and reduces progesterone secretion, leading to a longer calving index. Cystic ovarian disease is more common. Other disease are also more likely to occur, particularly ketosis, lameness and mastitis. Two indicators that the herdsperson can use to assess the energy status of early lactation cows are the condition score change and the ratio of fat to protein in the milk. Cows with low feed intakes postpartum have low milk protein contents and, if the ratio of milk fat:protein exceeds 1.5 for Holstein–Friesian cows in early lactation, there is an increased risk of these diseases. Cows fed a ration with 13–14% crude protein of low rumen degradability during the dry period are less likely to mobilize body protein to support the growth of the fetus than cows fed a ration with low crude protein or high protein degradability. They can, therefore, preserve body protein, which can then be used to maintain milk protein content during early lactation. However, if a ration with more than 13–14% crude protein is offered, the energy cost of detoxifying surplus plasma ammonia by converting it into urea in the liver is significant and wasteful. Ideally, a forage with low protein and energy density should be fed with a high undegraded dietary protein supplement. This will allow the cow to conserve body protein reserves without getting too fat, which would be likely if a high-protein/high-energy forage was fed, leading to rapid microbial growth and fat accretion in the body stores. Unfortunately autumn-calving cows are dry in the late summer period, and in the UK they are often kept on pasture which is quite high in energy and degradable protein content. Intake can only be restricted by stocking them at a high rate.

The importance of feeding additional concentrates to cows in the dry period depends to some extent on the post-calving diet that will be offered to the cows. If they are to be offered a high-energy diet, perhaps as a complete diet, little or no benefit in milk production will be obtained from additional feeding in the dry period. If they are to be fed a diet restricted in energy and protein content during the lactation, additional feed offered during the dry period will help to build up reserves that will be useful in early lactation, but it will make the cows more susceptible to metabolic diseases. Thus early research work, when cows were fed on restricted rations during lactation, found that additional dry-period feeding increased milk yield. Later research work, however, when cows were fed better quality rations during lactation, did not. An overriding principle is that it is biologically more efficient to produce milk directly from feed, rather than via stored body tissue, but this may not fit in well with feed availability, the feed, for example, being more readily available and of better quality in the dry period of summer-calving cows at pasture than in early lactation.

Normally the dry period is 7–10 weeks in duration, and it is preferable to feed a high-fibre ration for the first 4–7 weeks and to increase the energy content just for the last 3 weeks. This should allow the body condition score to increase to about 3.5 at calving on the 5-point score described in Chapter 8. If it exceeds score 4 then disease problems are more likely to occur.

Calcium intake requires careful management before and immediately after calving. During early lactation the output of calcium increases dramatically because of the high content of calcium in milk, and this may be more than the cow can provide from body stores in the bone tissue. Cows can regulate calcium absorption and limit excretion by the production of parathyroid hormone from the parathyroid glands. If it is possible to restrict calcium intake before calving to approximately 3 g kg^{-1} DM, the increased activity of the parathyroid hormone will increase the absorption and reduce excretion of calcium from the gastrointestinal tract. This gives the periparturient cow a greater ability to conserve calcium stores during the critical early lactation period. When non-lactating cows are at pasture, the calcium intake will often be much higher than this, so parturient hypocalcaemia continues to be a problem in grazing dairy cow systems.

A second difficulty in balancing the ration of non-lactating cows occurs with the use of white fishmeal as a supplement, which is suitable as a source of UDP to preserve protein stores. However, it also has a high calcium content, and is therefore not suitable in herds with a significant incidence of hypocalcaemia.

The problem of hypocalcaemia is exacerbated by the reduction in feed intake at calving, which reduces calcium absorption to a low level, and by the high milk yields that the modern dairy cow produces in the very early stages of lactation. The problem is extremely rare in beef cows, which give a much lower yield than dairy cows even in the first few days of lactation.

Feeding during lactation

In many countries the output of milk from each farm is governed by a quota, so accurate rationing is essential if the farm is to match its milk production to the permitted sales of milk to a processor or retailer. Much research over the past 20 years has been directed at improving the predictability of milk production by the cow. For example, given a knowledge of the feed quality and intake, the extent of the cow's nutrient reserves that can be used to support milk production can now be estimated (Alderman and Cottrill, 1993).

The lactation can be divided into three periods (Fig. 2.1). The first is characterized by a rapid increase in milk production to a peak. The increase is more pronounced than in other mammals, or even beef cows, and it is this early lactation period when most metabolic and some infectious diseases occur. Taking this into consideration, a future priority for dairy cow breeding will be to develop cows that have a flatter lactation curve, i.e. they do not ascend to such a high peak milk yield and a high output is maintained for a longer period of time.

The deficit between the nutrients required for milk production and the nutrients available from feed consumption is met by the mobilization of body fat reserves and to some extent body protein and mineral stores. The nutrient deficits are caused by the failure of feed intake to increase as rapidly as milk

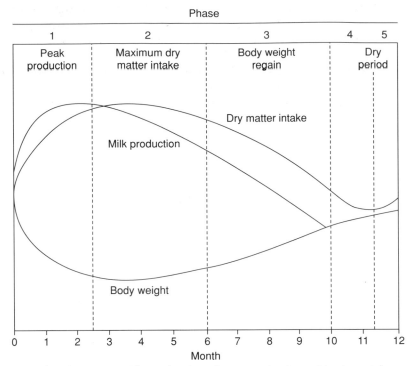

Phase

| 1 | 2 | 3 | 4 | 5 |

Peak production | Maximum dry matter intake | Body weight regain | Dry period

Dry matter intake

Milk production

Body weight

0 1 2 3 4 5 6 7 8 9 10 11 12

Month

Fig. 2.1. The changes in milk production, dry matter intake and body weight over the year of a typical dairy cow.

production in early lactation, so that it is not until mid-lactation that the energy balance is normally restored. This is in the second period, following gut involution, which is when peak DM intake is attained. A high-energy diet will accelerate the return to maximum intake, which is one advantage of allocating more concentrates to the early lactation period. An early return to maximum DM intake will advance the nadir of live weight, that should be reached during this second period. An excessive weight loss during this period will reduce milk yield, reduce the chances of conceiving and maintaining a viable embryo, and increase the risk of acidosis. If cows have a low level of body reserves, it is less likely that they will be able to endure a period of underfeeding without milk yield being reduced. If forage has to be restricted towards the end of winter because of inadequate silage supplies, both the expected duration and severity of the restriction should be taken into careful consideration when deciding whether to purchase additional feeds.

In the third period the loss of body tissue, which can be up to 80 kg in a high-yielding cow, should be regained, probably at about 0.5–0.75 kg day^{-1}. The decline in milk yield, which has been at a rate of about 2.5% per week since peak milk yield, continues and will accelerate if the cow is pregnant.

This is the idealized cow lactation, but in reality each cow is different. First lactation cows have a flatter lactation curve (Fig. 2.2), because they do not initially have the same milk production potential as cows, and in addition they have to divert nutrients to weight gain. Hence the increase up to peak lactation is particularly pronounced for high-yielding, older cows, compared with cows in their first or second lactation, whose lactation is preserved for a longer period. Total milk yield increases at least until the fourth or fifth parity (Fig. 2.3), and probably reductions in yield after this time are caused by an increased disease incidence rather than senescence. There are many examples of cows continuing to give high milk yields well into their late teens.

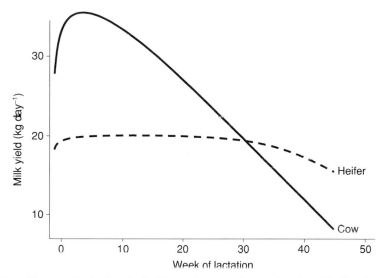

Fig. 2.2. Changes in the level of milk production of cows (——) and heifers (- - - -) over the period of lactation.

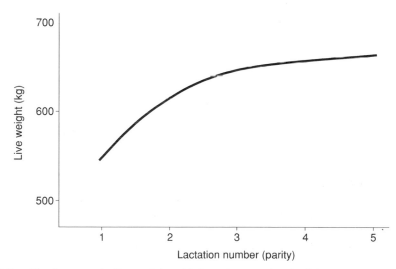

Fig. 2.3. The increase in live weight with lactation number in dairy cows.

Nutrition and milk composition

In most situations farmers are paid for milk by the volume, the content of nutrients that are of value to humans and its cleanliness. All of these are affected by nutrition and the effects of nutrition on milk yield and quality are intrinsically linked to each other.

Milk fat

Milk fat is produced both by the synthesis of fatty acids in the mammary gland and the absorption and secretion into milk of dietary fatty acids, with the bacterial digestion of feeds contributing to modification of the fatty acid profile in feed before it is secreted as milk or stored in the body (Table 2.7).

Acetic acid is the main precursor for milk fat synthesis, and as the acetogenic bacteria digest plant cell wall, the fibre content of the diet is the most important determinant of the fat content of the milk. The ratio of lipogenic nutrients (acetic acid, butyric acid and long-chain fatty acids) to glucogenic nutrients (propionic acid, glucose and some amino acids) therefore determines milk fat content, in particular the acetate to propionate ratio (Fig. 2.4). Fibre digestion is impaired if the rumen pH is less than 6.3, with acid-detergent fibre digestibility being reduced by about 4% per 0.1 unit reduction in rumen pH. Therefore, feeding large quantities of concentrates that are rapidly digested in the rumen to acid end-products should be avoided. Rumen pH can be maintained by feeding alkali treated forage or grain, or by stimulating saliva production, which contains sodium- and potassium-based buffers. It must be remembered that rumen pH is largely determined by the rate and composition of volatile fatty acids and saliva, not the feed. A high-yielding dairy cow, producing 300 l day^{-1} of saliva, will add more than 3 kg of sodium bicarbonate and 1 kg disodium phosphate to the rumen daily.

The buffering capacity of the rumen solids can also be important, this relates to the amount of acid or alkali that has to be added to alter the pH. When the rumen is subjected to an acid challenge after heavy concentrate feeding, there is less possibility of milk fat content being reduced if the rumen contents are able to absorb the acid production without the rumen liquor pH

Table 2.7. Concentration of the major fatty acids in lipids from fresh grass, milk and adipose tissue.

Fatty acid	Carbon chain number: number of double bonds	Grass	Milk	Meat
		\multicolumn{3}{c}{(g kg^{-1} lipid)}		
Myristic	14:0	10	120	30
Palmitic	16:0	110	310	260
Stearic	18:0	20	110	140
Oleic	18:1	50	240	470
Linoleic	18:2	120	30	30
Linolenic	18:3	620	10	10

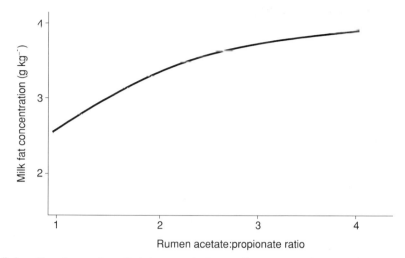

Fig. 2.4. The change in milk fat concentration with rumen acetate:propionate ratio.

varying too much. The requirement for rumen buffers is determined by the rumen pH and the buffering capacity of the rumen contents. Strictly speaking a rumen *buffer* will bring the pH up to the required level (6–7) but not beyond. Some alkalis, such as magnesium oxide, are good acid-consumers and alkalinizing agents, and will increase rumen pH above this level. They are not buffers. Other potentially good alkalinizing agents, such as limestone, do not dissolve at normal rumen pH. Sodium bicarbonate is the most commonly used rumen buffer. Sodium chloride is pH neutral but can still increase rumen pH by increasing the amount of sodium buffers in saliva. Sodium is returned to the gastrointestinal tract via saliva, which means that increasing the sodium content of feeds increases rumen pH even if the sodium compounds in the plant are pH neutral.

Milk fat content starts to decline if the proportion of forage in the diet is less than 40%. The reduction becomes increasingly severe as the forage content declines, and if a diet of only 10% forage is offered, the milk fat content is likely to be only 20 g kg^{-1}. Forage is an imprecise term, so a direct measurement of the digestible fibre content of a diet will relate more accurately to its potential for milk fat production. The fat content starts to decline when the diet contains less than about 300 g neutral-detergent fibre kg^{-1} DM. Two cases where forage intake might appear adequate but can lead to low-fat milk are, firstly, where cows are grazing spring grass and, secondly, where the forage has been ground or comminuted to a particle length of less than approximately 0.6–0.7 cm. A total mixed ration is an effective way of avoiding excessive acid production in the rumen and low-milk fat content, as the concentrate is consumed with the forage and the release of nutrients synchronized.

Feeding concentrates frequently, perhaps four to six times per day through an out-of-parlour feeder, rather than twice per day in the parlour will help to prevent milk fat depression (sometimes referred to as the Low Milk Fat

Syndrome). However, if the diet is so low in fibre that milk fat production is severely reduced, this cannot be fully rectified by more frequent concentrate feeding and the solution should be to change the diet rather than increase the frequency of feeding concentrates. If a high-concentrate intake is required, it is best to use ingredients with a high content of digestible fibre, such as sugarbeet pulp, rather than a concentrate high in starch, such as cereals.

Low-energy diets also restrict milk fat synthesis, and high-protein diets reduce milk fat content by increasing milk yield, emphasizing that the effects on composition cannot be considered in isolation from the effects on milk yield. The incorporation of unsaturated fatty acids into milk is possible if fats are protected from rumen fermentation. Normally no more than about 6% of unprotected fat should be included in the diet as it has adverse effects on rumen digestion:

- physical coating of the fibre preventing microbial attack;
- toxic effects on some microorganisms, thus modifying the rumen microbial population;
- inhibition of microbial activity from surface-active effects of fatty acids on cell membranes;
- reduced cation availability from formation of insoluble complexes with long-chain fatty acids.

Fats are more toxic to rumen microbes when they are unsaturated, but if fed as whole oil seeds to high-yielding dairy cows, whose rumen turnover rate is high, the adverse effects on rumen digestion can be minimized. However, if fat is protected by being complexed with calcium as a saponified product or with formaldehyde, it can act as a source of additional energy to supplement the fermentable metabolizable energy (FME) and increase the unsaturated fatty acid content of milk fat. The lipid–calcium or lipid–protein complex is insoluble and hence undegradable at the normal rumen pH (6–7) but degrades rapidly in the acid conditions of the abomasum (pH 2–3). If unprotected unsaturated fats are fed, they are mostly hydrogenated in the rumen. The feeding of calcium soaps of unsaturated fatty acids will only be successful if rumen pH remains consistently above 6; often in rations for high-yielding dairy cows, the high concentration of rapidly fermented starch reduces rumen pH below this level, which results in dissociation of the calcium soap. The addition of dietary buffers may be necessary.

The incorporation of unsaturated fatty acids into milk may reduce the keeping quality of milk as they can be rapidly oxidized (Goering *et al.*, 1976). Free fatty acids, in particular butyric acid resulting from lipolysis, then give the milk a rancid flavour. The addition of an antioxidant, such as alpha-tocopheryl acetate, will lengthen the keeping life of the milk, but in Europe at least milk cannot have substances added to it before sale, and elsewhere there may be consumer resistance to the addition of antioxidants to milk. In butter, increasing the content of unsaturated fatty acids makes it spread better at low temperatures.

Milk fat is naturally palatable, but it contributes to the high fat intake that is associated with problems of atherosclerosis and heart disease in the developed world. As a result, where lifestyles tend to be sedentary, consumers have switched to consumption of milk that has had either most (skimmed milk) or some (semi-skimmed milk) of the fat removed. This does little to reduce the palatability of the milk; in fact most consumers would find it difficult to even distinguish semi-skimmed and whole milk by taste alone. However, the colour of whole milk is more yellow than skimmed or semi-skimmed, because of the increased concentration of carotene and its derivatives, and this will indicate whether the milk has been skimmed or not. The farmer is paid less for milk fat than protein, and in some countries is subject to a quota on milk fat production, so there is little benefit to producing milk with very high fat concentrations. This is unfortunate as the potential to modify milk fat content is greater than other constituents, largely because the genetic variation and the impact of nutrition is greater.

Milk protein

Milk protein is mainly casein, about 70%, with the remaining 30% comprising beta-lactoglobulin, alpha-lactalbumin and immunoglobulins. It is the casein that is of particular value for the clotting of cheese and yoghurt because it precipitates in mild acid. Milk protein has a naturally high content of lysine, which complements cereal protein well, and is one of the most economic forms of animal protein available. A small proportion of people are allergic to milk protein, but the effects are transitory and diminish with age. Although milk protein is present in milk at lower concentrations than fat or lactose, it is in many countries the most valuable constituent of milk because of its necessity for the production of cheese and yoghurt.

The natural variation in milk protein content is considerably less than for fat, so the opportunity to select cattle with increased milk protein is limited. In addition, if milk protein content is selected for, milk yield will decline, so genetic progress in increasing milk protein yield is slow. Variation in milk protein concentration to changes in nutrition are less than milk fat too, but a clear relationship between the energy supply to the cows and milk protein content has been established. Energy intake can be increased by increasing the concentrate intake or increasing forage quality. The response in milk protein content is mainly caused by the release of amino acids from being deaminated to supply energy. Cows that are catabolizing body tissue to provide for their energy requirements, therefore, tend to have low milk protein concentrations, as some of the feed protein will be utilized for energy. For this reason it is preferable for cows to calve at a condition score of no more than 3, so that their feed intakes are high post-calving and milk protein content will then be acceptable (at least 3–3.5 g kg^{-1}).

Typically an increase of 10 MJ of ME intake increases milk protein content by about 0.6 g kg^{-1}. However, the response is curvilinear, with smaller responses in milk protein content at high-energy intakes. The increase in milk protein content is accompanied by an increase in milk yield, making it difficult to increase milk revenue by this method if milk sales are restricted by a quota. In some cases high yields lead to low milk proteins and the feeding of fat supplements can reduce milk protein content by up to 3 g kg^{-1}.

Cows often have to mobilize body protein during the dry period to sustain the growth of their calf. This can be avoided by feeding a protein supplement of low rumen degradability in the non-lactating period as described previously, leaving body protein available to support the protein requirements for milk production. Additional crude protein fed during lactation will also increase milk protein; if the ratio of protein to energy intake is thereby optimized for the rumen bacteria, enabling feed intake to increase. If the ratio of crude protein to ME in the diet is high, surplus ammonia will be converted into urea and some will pass into the milk to increase the milk non-protein nitrogen content. This is of little value to cheese manufacturers, even though farmers will be paid for it.

Specific amino acids may be inadequate for the maximum growth of the rumen microorganisms in a high-yielding cow, particularly if she is fed a maize silage-based diet. Supplementation with amino acids that have been protected from rumen degradation in the same way as unsaturated fatty acid supplements, can give economic responses in milk yield and increased milk protein content. However, responses are variable and it is difficult to predict which, if any, amino acids will be in short supply. The essential amino acid, methionine, is most likely to be deficient in high-yielding cows. If they are fed a maize silage-based diet lysine may be co-limiting. The non-essential amino acid, glutamine, constitutes 25–30% of the major milk protein, casein. It is also glucogenic and is deficient for longer than other amino acids at the start of lactation. The optimization of the amino acid content of the diet could assume considerable importance if the control of nitrogen emissions from cattle becomes more urgent. Feeding the correct amino acid balance for maximum microbial growth and in the dietary undegraded protein will be important, rather than assuming that the excess nitrogenous compounds can be excreted or partially recycled. Typical essential amino acid contents in the major cattle feeds are shown in Table 2.8.

Table 2.8. Concentrations of some essential amino acids (% of crude protein) in typical cattle feeds.

	Soybean meal	Maize gluten feed	Maize gluten meal	Dried distillers' grains	Dried brewers' grains	Fish-meal	Barley	Maize silage	Grass silage
Methionine	1.5	2.4	3.2	1.9	2.2	2.9	1.6	0.9	1.2
Lysine	6.6	2.9	1.7	2.2	3.2	7.6	4.6	1.8	3.4
Isoleucine	5.7	4.3	3.8	3.7	7.2	4.6	3.7	2.8	4.7
Valine	5.5	5.0	4.5	4.8	6.1	5.5	5.4	3.7	6.0
Leucine	7.7	9.1	15.7	11.1	11.5	8.1	7.0	6.5	3.5

Milk lactose

Milk lactose is made of two simple sugars, glucose and galactose. It is syn
thesized in the Golgi apparatus and secreted into the milk along with protein.
It is the major osmotic regulator in milk, although chloride, sodium and
potassium also play a part. As a result, when mastitis increases the sodium
content of milk caused by the junctions between the alveolar cells becoming
less tightly bound, the lactose content of the milk is reduced by about 2 g kg^{-1}.
Lactose content in milk may decline by 1–2 g kg^{-1} when cows are underfed in
late lactation, perhaps as a result of the sodium content increasing more at this
time.

Milk lactose is of little value to the processor, at least in the UK, because it
is in surplus. As such, it is interesting that breeds with low yields of milk with
high fat and protein contents, such as Channel Island breeds, use less energy
on lactose output than breeds with high yields of low fat and protein content,
such as the Holstein, since the lactose content of the milk from these two
breed types is similar. Ironically it is one of the milk consitituents valued by
consumers, as sweetness is one of the main determinants of milk acceptability.

Minerals and vitamins

The mineral content of milk is important for human nutrition, with calcium,
iron and zinc being of particular importance. The concentration of many
minerals reflect blood plasma concentrations and often dietary status, but the
epithelial cells can act as a barrier or may transport the minerals after
complexing with organic compounds. Most calcium is organically bound, with
casein, phosphate or citrate. The concentration of iron, zinc and copper are
closely regulated and plasma concentrations do not affect milk concentrations.
Hence blood haemoglobin and even ferritin concentrations do not influence
milk iron concentrations. Milk sodium concentration is also homeostatically
regulated, but a severe deficiency can reduce milk sodium concentrations, as
can occur with zinc. By contrast for selenium, also of major importance for
human nutrition, there is no evidence of regulation and milk concentrations
reflect those in the plasma.

Vitamins are transferred unchanged from the blood to the milk, so
increases in vitamin status are reflected in higher milk vitamin concentrations,
particularly the fat-soluble vitamins. The most important vitamins in milk are
A, B$_2$ and B$_{12}$. Reducing the fat content of milk reduces the concentration of A,
but not B$_2$ or B$_{12}$.

Stage of lactation effects

Changes in milk composition over the cow's lactation reflect the changes in
yield, energy balance and feeding level. In the first few weeks, milk fat content
declines rapidly as yield increases (Fig. 2.5). After week 4, milk fat content
gradually increases for the rest of lactation. Milk protein content declines
gradually over the first 12 weeks of lactation, as the cow mobilizes body tissue
to sustain lactation. Thereafter it increases again, until by the end of lactation it
is back to approximately where it started. Milk lactose content initially

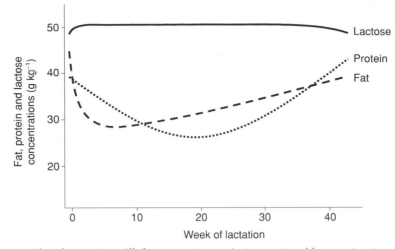

Fig. 2.5.   The changes in milk fat (– – – –), protein (- - - - -) and lactose (——) concentrations over the lactation of the cow.

increases as colostrum is replaced by milk. Thereafter it changes very little over the lactation, except that it may decline towards the end of lactation, particularly if the cows are underfed at this time. It is difficult to distinguish true effects of stage of lactation from the changes that take place in milk yield, appetite and body condition that occur as the lactation progresses. Two of the major milk constituents, fat and protein, reflect the change in milk yield that occurs during the lactation, suggesting that some of the 'lactation effects' are in fact caused by dilution. The absence of variation in milk lactose concentration over the lactation may reflect its role as an osmotic regulator.

Taints in the milk from feeding

Milk is likely to acquire taints when it is in an 'open' system, i.e. exposed to the environment, particularly if it is collected into cans in the feeding or living area of the cowshed and transferred to churns. Brassicas and beets may produce a taint if fed within 3 h of milking, so should preferably be fed after milking. Other strong smelling plants, such as wild onion or stinking mayweed, can create a taint in milk if cows are short of grass and browse around the hedgerows. Feeding practices that lead to acetonaemia should be avoided, as this can taint the milk with a characteristic pear-drop taste.

The quality and stability of milk produced from by-product feeds may differ from that derived from feeds used in routine dairy cow feeding. For example more traditional by-products such as brewer's grains have high concentrations of unsaturated fatty acids and are expected to have a high level of *trans*-fatty acids, which affects the stability of milk by making it more susceptible to oxidation. High levels of transition metals in feeds can promote this oxidation through decomposition of lipid hydroperoxides. The maximum content of metal pollutants in feeds, particularly industrial by-products, has therefore to be carefully considered.

In developing countries, the distribution and storage conditions for feeds are important factors affecting milk quality, especially under the extreme temperature conditions found in the tropics. The composition of the raw milk is also important in order to minimize undesirable flavour components and oxidation. Another factor affecting milk quality is the transfer of flavour components from by-products to the milk, which can give an undesirable off-flavour. Metals can precipitate oxidation as described previously, but anti-oxidants such as vitamin E can retard the reaction. In Europe, milk cannot legally have substances added to it before sale, but this may be desirable in developing countries. Finally, some metal containers for milk storage may add undesirable metallic taints to the milk!

Foods Suitable for the Dairy Cow

Forages

Silage

In most countries, grass or other feeds for grazing cannot grow all year, so farmers conserve surpluses from times when growing conditions are good so that they can be fed to the cattle during the winter months (when it is too cold for grass to grow) or in the dry season (when there is insufficient moisture). In temperate countries, most fodder is conserved as silage, which is cut at a younger stage of growth than hay and, therefore, tends to be more nutritious. In the humid tropics, making good quality silage is more difficult because the high temperature and humidity make controlling the fermentation more difficult and because it requires more equipment and facilities than hay-making. Nevertheless, the intensification of cattle production in some tropical regions has led to a resurgence of interest in silage as a means of providing fodder in the dry season.

The digestibility of silage in the cow's gastrointestinal tract is one of the most important factors influencing its value to the dairy cow for milk production. It is mainly determined by the age of the grass at cutting, but a rapid fermentation, with an additive if necessary, and efficient sealing of the clamp are also important in producing grass silage of high digestibility. Additives will be required if the silage is wet, not finely chopped and has a low content of sugars to sustain a rapid anaerobic fermentation. There are three main types of additives used – soluble sugars, inoculants (both of which act by encouraging the bacterial fermentation), and acids, which simulate the end-products of fermentation to prevent further degradation of the grass. Acid additives may be combined with the sterilant, formaldehyde, to ensure further fermentation is prevented.

The preferred type of fermentation is by lactic acid producing bacteria, which cause a rapid reduction in pH and stable conditions in the ensiled material. If air is not squeezed out of the silage after it has been put into the clamp and if the sealing of the clamp is not sufficient to exclude air, then clostridial bacteria may multiply. The acid produced – butyric acid – is weaker than lactic acid and fermentation will continue for longer with more extensive breakdown of protein. Usually about 40% of the crude protein is true protein, but it is the composition of the non-protein nitrogen that is the best indicator of silage fermentation quality. In a well-made silage, only 5–12% of the non-protein nitrogen is in the form of ammonia; in a 'butyric' silage this may be 20–30%, and thus the proportion of non-protein nitrogen that is ammonia is used as an indicator of silage quality.

Once the clamp is opened for feeding and is exposed to air, there is a further risk of bacteria or fungi fermenting the crop, with associated loss of digestible material and in some cases a health risk to the cattle consuming the silage. A tower silo prevents exposure of the silage to the weather during feeding, but at a much greater cost than silage 'clamped' between walls made of concrete or wood. A clamp for silage usually has a concrete floor, which is hardwearing but the concrete will eventually be eroded by the acids in the effluent. An earth floor allows the effluent to seep away but the silage fed to the cattle may be contaminated. Only high-DM silages can be stored in tower silos and there may be significant field losses if the grass has to be dried slowly in damp conditions in the field.

Silage crops may be cut up to four times each summer in Britain, but more normally only two cuts are taken, which provides a fresh area for grazing, or aftermath, at the end of the grazing season. This allows stocking rates to be relaxed at this time, when grass growth is slow. The more frequent the cutting, the greater the digestibility of the silage, since the grass will be cut at a younger stage of growth. When considered over the same period of the grazing season, a frequent cutting system provides flexibility in the availability of aftermaths, but reduced yields of high-quality silage (Table 2.9). This will

Table 2.9.   A comparison of grass growth, dairy cow performance and profitability with three- and two-cut systems of silage conservation.

	Two-cut	Three-cut
Herbage yield[a] (t DM ha^{-1})	9.3	7.8
Herbage metabolizable energy (ME) content (MJ kg^{-1} DM)	9.3	11.1
Utilized ME from herbage (GJ ha^{-1})	91	84
Silage DM intake (kg per cow)	9.3	10.1
Milk yield (kg per cow)	19.5	20.7
Live weight gain (kg day^{-1})	0.24	0.36
Cow feeding days ha^{-1}	1020	800
Milk yield (kg ha^{-1})	19707	16457
Relative margin[b] per cow	83	100
Relative margin[b] ha^{-1}	107	100

[a]After field and storage losses have been deducted.
[b]After purchased feed has been deducted.

often lead to greater milk output and profit margin per cow, but the lower yields and increased intake per cow can lead to considerably reduced cow feeding days. The greater yield of ME per hectare from a less frequent cutting system could increase the profit margin per hectare.

A frequent cutting system, which produces less silage but of high digestibility, will provide fewer cow feeding days. It will, however, sustain high milk yields per cow as long as it lasts, whereas cows fed silage cut less frequently will require more concentrate supplement to produce the same amount of milk. This may reduce the profitability per cow. An increase in the digestible organic matter concentration in the DM ('D' value) of 10 g kg^{-1} has been found to increase silage DM intake by approximately 0.25 kg and milk yield by 1.3 kg. Even though 0.25 kg silage DM only has an ME content of about 4 MJ, which would be expected to provide the energy for less than 1 kg milk, the increased digestibility of high-quality silage will provide the nutrients for an even greater increase in milk yield.

It is not just the digestibility of the silage that indicates whether the cows will support high milk yields. Quality is also indicated by the fermentation characteristics. Silage DM content may be insufficient for high-DM intakes. If the DM content is less than about 180 g kg^{-1}, DM intake is reduced and effluent losses from the clamp will be considerable (Vérité and Journet, 1970). Wilting grass before it is ensiled will help to reduce the effluent production and the weight of material that has to be transported to the silo. Cows fed wilted silage consume more DM but do not necessarily produce more milk. The quality of the fermentation of wilted silage is often greater than if the crop is cut and directly taken to the silo (direct cut silage), and if this is so milk yields are likely to be increased by wilting.

Grass cut for silage may be chopped to varying degrees by the harvesting machine (single-chop, double-chop and precision- or fine-chopped) before ensiling. The intake of the shortest material, the precision-chopped silage, contains particles about 1 cm in length. This is considerably shorter than silage harvested with a single chop, which will have particles at least 5–10 cm in length. The fermentation is usually better with precision-chopped material, as it is easier to exclude air by rolling with a tractor after it has been put into the clamp. Precision-chopped silage, however, can be difficult for cows to remove from a self-feed clamp, especially if they are heifers that are losing their milk teeth.

Maize silage is popular with cattle farmers because it has several advantages over grass silage. Slurry can be applied in large quantities before planting and much less nitrogen is required compared with grass. The crop will usually produce good yields of highly digestible cattle food after only one harvest, at the end of the growing season, whereas grass needs to be harvested two or three times. The water requirements of the maize crop are high, limiting its geographical range. The same forage harvester can be used for harvesting as for grass silage but with a modified pick-up attachment. The ME content of maize silage is usually greater than grass silage but the crude protein content is less (Table 2.10). If the maize is left to become mature before

Table 2.10. The composition of typical maize and grass silages.

	Grass silage	Maize silage
Dry matter (g kg^{-1})	18–30	25–35
Metabolizable energy (MJ kg^{-1} DM)	10	11
Crude protein (g kg^{-1} DM)	17	10
Crude fibre (g kg^{-1} DM)	30	20

harvesting, more of the energy is in the cob and the rest of the plant is of low digestibility.

In tropical countries, the high temperatures allow the crop to rapidly grow to maturity. Here the cobs are usually harvested for human consumption or occasionally to provide animal feed supplements, leaving a fodder residue which can be either chopped and fed direct to cattle or grazed *in situ* by the cattle during the dry season. It does not support high growth rates in beef cattle nor provide sufficient energy for high milk yields, but it will sustain the cattle during the dry season when there is often little other fodder available.

In temperate regions, maize for silage is usually harvested at an earlier stage of growth during the autumn. A good compromise between yield and quality is provided by harvesting when the grain is at the medium to hard dough stage, and the DM content of the crop is about 300 g kg^{-1}. Earlier maturing varieties developed in recent years now allow maize silage to be grown in more extreme latitudes.

Feeding silage

The silage intake of groups of cows must be monitored so that mean individual intakes can be calculated. In intensive systems the silage is usually fed *ad libitum* so that cows always have some available. In traditional feeding systems in byres, it was possible to monitor individual intakes, however, most cattle are now fed in loose-housed groups. After the silage has been made, the farmer must assess the weight of silage conserved to determine whether adequate silage is available to meet cattle intake requirements for the period of housing. If the silage is clamped, this can be determined from a knowledge of the size of the clamp and the density of silage within it.

Silage can be fed either directly from a clamp, so-called self-feeding, or it can be extracted from the clamp by machine and fed along a passageway or in a circular feeder. If the silage is fed in a passageway, cows should be restrained behind a barrier that allows them to put their heads through to feed, but not to walk on the silage or pull their heads back through the barrier while they are still eating silage. This is usually achieved by having a tombstone or diagonal bar configuration on the upper half of the barrier (see Fig. 7.3). Cows must lift their head before they can withdraw them from a tombstone barrier, and with diagonal bars they must twist their heads.

Either a block-cutting machine can be used to extract the silage from the clamp in cuboid shaped blocks or it can be teased from the clamp by a fore-end loader, in which case air will enter both the extracted silage and the clamp, accelerating secondary fermentation in warm conditions. A disadvantage of self-feeding is that the clamp can only be made up to a (settled) height of about 1.8 m because of the limited reach of the cattle, depending on their size. It may be possible for it to be made taller, and the top layers removed with a block-cutter and placed in a feeder or along a feeding barrier. If the cattle cannot reach the silage at the top of the clamp, this may fall down and may be trampled by the cattle. Cattle can be prevented from reaching too far into the clamp by an electrified bar at about 0.9 m from the floor, suspended from a length of angle-iron driven into the silage, or by an iron pipe supported on a platform that the cows stand on. The rate at which these barriers move forward determines how much silage the cattle are allowed, and so should be accurately monitored. The angle iron should be driven further into the silage at daily or twice daily intervals, thus making fresh silage accessible.

Sometimes young cattle are timid about feeding, particularly if an electrified bar is used, and will visit the feed face mainly at night. Intake may be reduced by 10% with an electrified pipe, but wastage is likely to be less. To prevent any constraint on young cattle visiting the silage feeding area, the face should be sufficiently wide to provide at least 150 mm of face width per cow. It should be lit throughout the night, otherwise this acts as a disincentive for the cattle to visit. The advantages of self-feed silage are the low cost and the ability to keep the cows occupied in what is essentially a system of 'vertical grazing'. In passageway feeding systems, aggression and other deleterious behaviours can be stimulated by the short time spent feeding (approximately 6 h, compared with 7–8 h for self-feeding and 9–11 h for grazing) and the lack of comfortable lying areas. Any advantage of reduced labour and machinery for feeding may be offset by having a larger area to clean, which also produces more dirty water that has to be disposed of safely.

Passageway feeding is used by many to achieve high intakes, especially since it is suitable for combining silage and concentrates together in a complete diet. A forage wagon can be used to feed just silage or to create a simple mixture of feeds. Complete diets are made up in a mixer wagon, which is described later in this chapter. The forage wagon meters out the feed alongside a feeding barrier, and a simple mixed diet can be made by layering concentrates and silage on top of each other, but it will not be well enough mixed to ensure that a diet of equal concentrate:forage ratio is consumed by all cows. The simplicity is, however, attractive. Some ingredients, such as rumen-undegraded protein or mineral/vitamin supplements need to be accurately rationed to the cows, otherwise they may cause a metabolic disturbance. These might be rationed to the cows through individual feeders either out of, or in, the parlour.

A further option for feeding silage that is increasingly popular on small farms is to wrap silage bales with plastic, three or four layers, or place it in a bag. The silage made in this way loses less effluent and the fermentation is restricted. It can be offered in a circular feeder on a concrete standing or in a sacrifice field. This method of feeding silage can also be used to supplement grazing cattle, and the feeder moved when damage to the sward occurs.

Hay

Hay relies on preserving grass by removing the moisture that microorganisms require for survival. This is usually done by drying the crop in the sun, requiring up to 5 days with regular turning by mechanical or manual means to accelerate desiccation. This laborious process has resulted in haymaking declining in popularity in many intensive European cattle farms because of improvements in silage-making machinery and increases in herd size. Silage is more suitable for feeding to large numbers of cattle as machinery is available that can conserve and feed large quantities of silage rapidly. Hay was tradition-ally made in small bales, of about 20 kg, which could be handled by one person; it can now also be made into larger bales, of 600–900 kg, which are handled mechanically. Hay is usually made in one late harvest since the grass declines in moisture content as it matures. This harvest is usually taken about one half of the way through the grazing season. Excessively long field-drying has the risk of leaf loss, and the resulting material is inevitably of low digest-ibility, which is another major reason for many farmers turning from making hay to silage. Energy losses from the grass plant are often very high during haymaking, because of continued plant respiration. However, if the same grass could be used to make either hay or silage, the protein value of the hay would usually be greater than silage, because there is less protein denaturation during the conservation process.

Grass can be artificially dried in a barn with forced air. This reduces energy losses in the plant, but it is prohibitively expensive because of the energy requirements for drying. The end-product is of better quality than field-dried hay, and safer, because hay is often baled before it is properly dry and moulds form in storage. Mouldy hay causes an illness called Farmer's Lung, which is an immune complex hypersensitivity, and abortion in cattle. When people handling hay get Farmer's Lung, antigens stimulate antibody production by the immune system, which react with further antigens to form immune complexes. The immune complexes activate complement and attract phagocytes, which release lysosomal enzymes, causing tissue damage. To avoid this, hay should contain less than 17% moisture in storage, and should not be cut in the field if the moisture content is greater than 25%.

Straw

In many parts of the world, straw or other crop residues, such as maize stover, are important feeds for cattle, particularly beef cattle. These are the stalks of plants that are left over after the grain has been removed for human consump-tion. The available energy content is low, as most of the energy is locked up in the form of cellulose and other structural carbohydrates that are lignified. Cattle will not consume much straw because its rate of breakdown in the rumen is slow. Hence if alternative roughages are available cattle usually prefer them to straw. The protein content of straw is much less than that required by most cattle, often only 40 g kg^{-1} DM. The content of minerals and vitamins is also low. For high yielding dairy cows straw, therefore, contains little nutritional value, other than occasions when the fibre is required to maintain rumen function and animal health (Frank, 1982). Most cows will consume some of the straw that is used as bedding in a straw yard, but it is unlikely to contribute much, if anything, to their energy requirements and straw cannot be included at more than 20% of the DM in the diet of highly productive cattle without their performance suffering. In particular, straw should not be fed in significant quantities to early lactation dairy cows. The energy value of straw depends on the rate of turnover of rumen contents. In a dairy cow, this may be reduced by the consumption of large quantities of material of low digestibility, and the bacterial growth will decline. The energy value of straw that is normally given in feed tables (6 MJ kg^{-1} DM) may be too high in these circumstances.

Straw can be upgraded by treating it with chemicals, especially alkali agents such as sodium hydroxide or ammonia (Jackson, 1977). These reduce the lignification of the structural carbohydrate, rendering it more available to breakdown by the rumen microorganisms. The economic viability of the process is questionable in many regions, and the availability of the chemicals for treatment is often a problem in developing countries, but new techniques of predigestion with microorganisms may surpass chemical treatment in both effectiveness and economic viability.

Straw can be harvested together with the cereal grain in the form of 'whole-crop' or arable silage. In theory harvesting costs can be reduced and high DM yields can be achieved, but there is a risk of significant losses during processing and/or storage. The system is less flexible than if the grain and straw are harvested separately. In addition, not all the grain will be utilized by the animal, because it is less processed than when it is harvested separately. Treatment with urea before ensiling will improve preservation.

Concentrated feeds

Concentrated feeds, or concentrates, are based on cereals or other high-energy and protein feeds. They are usually made into a pellet, or compound, with the addition of a binding agent, most often sugarcane molasses. They are brought onto the farm either in bags or they may be delivered by an auger into a feed bin. The cost of these processes makes compound pellets an expensive form of cattle food, unless the raw ingredients are inexpensive. The most important decision for a dairy farmer to make concerning cattle feeding is how much concentrated feeds to offer. Dairy cows can give yields of 7000 l per cow per lactation on high-quality forage alone, but in most situations, farmers get an economic response to providing at least a low level of supplementary concentrates. In developing countries, less concentrate supplements are fed than in the industrialized countries, and they are often of lower quality, because cereals are relatively expensive and are reserved mainly for feeding to humans.

In deciding whether to increase the concentrate feed on offer to the cows, a manager should estimate the anticipated increase in milk yield and determine a break-even point where additional concentrate feed would provide no additional financial return (Fig. 2.6). However, feeding additional concentrates will do more than simply increase milk yield, and the manager should consider the changes in milk composition and savings in forage when concentrate intake is increased. Usually, milk fat content will progressively decline and, to a lesser extent, protein content will increase, as the amount of concentrate and the ratio of glucogenic to lipogenic precursors increases. At most levels of concentrate intake, milk fat and protein *yields* increase with concentrate intake, but at very high intakes, in excess of approximately 12 kg day^{-1}, the milk fat content declines rapidly with additional concentrate, so that milk fat yield declines (the Low Milk Fat Syndrome). Under virtually all circumstances this will be an uneconomic level of concentrate to feed to cows.

The saving in forage when concentrate supplements are fed should enable the stocking rate of cattle on the farm to be increased. However, in dairy cows, this will increase milk production, which may cause the farm to exceed its milk quota, and alternative enterprises, such as a beef production unit, will have to be found to utilize some of the land. The dairy farm manager must decide how to maximize income to the farm, and this will entail looking at responses to the different resources on the farm, in particular the most limiting ones (Table 2.11). Research conducted by the West of Scotland College of Agriculture (Leaver and Fraser, 1987) identified the effects on farm profitability of feeding different levels of concentrates to Friesian dairy cows. Bearing in mind that the relative profitability depends on the prevailing economic situation, in particular the cost of the major resources in relation to the value of the output,

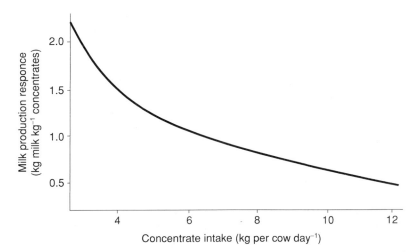

Fig. 2.6. The additional milk produced per kg concentrate fed for different levels of feeding a typical dairy cow.

Table 2.11. Annual performance and profitability of dairy cows in two 'farmlets' of the West of Scotland College of Agriculture, fed two contrasting levels of concentrates (Leaver and Fraser, 1987).

	Concentrate level	
	High	Low
Concentrate intake (t year^{-1} per cow)	2.2	1.0
Milk yield (kg per cow)	6000	5100
Milk fat content (g kg^{-1})	40.0	40.8
Milk protein content (g kg^{-1})	34.8	33.8
Relative gross margin per cow[a]	100	104
Stocking rate (cows ha^{-1})	2.61	2.15
Relative gross margin ha^{-1}	100	79
Relative gross margin l^{-1}	100	116

[a]After payment for purchased feeds.

feeding additional concentrates increases the gross margin per hectare (largely because stocking density increases) and reduces the gross margin per litre (because feed cost per litre increases). Under strict quota limitations, therefore, farmers may prefer to feed low levels of concentrates, but if the land area of a farm is restricted the farmer may prefer to feed more. The output per cow is likely to be particularly important to farmers keeping pedigree cattle, who may feed high concentrate levels so that the cows' performance is high and others are encouraged to buy them.

In traditional systems where cows are individually tethered in stalls, the forage allocation is usually restricted because it has to be delivered to each cow by hand. Hay is most often fed, rather than silage, and because it is usually of lower quality than silage, the forage will often provide less than the energy requirements for maintenance of the cow. In these circumstances increasing the concentrate allocation results in a major increase in milk yield because the additional concentrate acts as a true supplement, rather than substituting for forage. Potentially, each kilogram of concentrate is able to supply up to the energy requirements for 2.5 l of milk. However, not all the concentrate is used for milk production. The response in milk yield is greater from early lactation cows, because they allocate more of the nutrients to milk production and less to weight gain. In such systems it is therefore beneficial to allocate a disproportionate amount of the total concentrate ration for the lactation to cows in early lactation.

In modern loose-housing systems, forage is usually made available *ad libitum* because it can be fed by machine, and concentrates are fed individually, either in the parlour or outside. In this case, additional concentrates will substitute for some of the forage intake, and the total intake will not increase as much as when forage is fed at a restricted rate. There will be little or no benefit in offering extra concentrates to cows in early lactation, because the substitution rate is greater in early lactation than later on when appetite has increased.

The allocation of concentrates to dairy cows should take account of their physiological state, lactating or non-lactating, pregnant or non-pregnant. Most dairy farmers prefer to feed more concentrates to those cows giving the most milk whatever their forage feeding system, but a considerable amount of research with loose-housed cows fed high-quality forage *ad libitum* has demonstrated that responses are similar from cows that are producing different quantities of milk. Cows may produce different amounts of milk either because they have a different genetic potential for milk production or because they are in different stages of lactation. As nutrient intake increases, particularly energy, milk yield increases up to a certain level and then no further increase can be obtained from greater energy intakes (Fig. 2.7). The extra nutrients are stored as body tissue, to be used at a later date. If the responses of a high-and low-yielding cow to an increase in ME intake (from a to b, Fig. 2.8) are compared, they are the same from both cows (from c to d for the low-yielding cow and e to f for the high-yielding cow). However, at high-energy intakes an increase in energy supply (from g to h) may produce a response from the high-yielding cow (from i to j) but not from the low-yielding cow (point k).

The response of high-yielding cows to increased energy supply from concentrates may be offset by a similar reduction in low-yielding cows, if they receive fewer concentrates. However, it may be worthwhile allocating rumen-protected protein and mineral/vitamin supplements solely to cows with

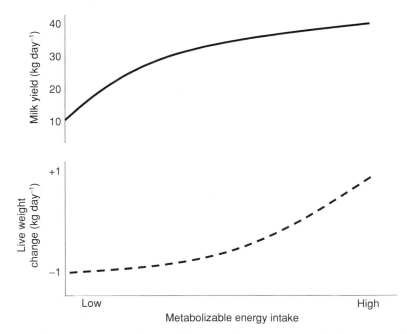

Fig. 2.7. The responses in milk yield (——) and live weight (- - - -) of a typical dairy cow to increasing metabolizable energy intake.

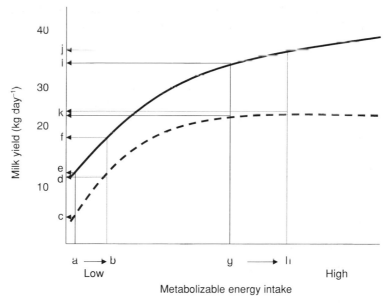

Fig. 2.8. The conceptualized responses of a high-yielding cow (——) and a low-yielding cow (- - - -) to increases in metabolizable energy (ME) intake, both at a low level of intake (point a → point b) and a high level of intake (point g → point h). Responses are indicated at the low level of intake by an increase in milk yield from point (c) to point (d) for the low-yielding cow and from point (e) to point (f) for the high-yielding cow. At the high level of intake, absence of response in the low-yielding cow is indicated at point (k); the high-yielding cow increases her yield from point (i) to point (j).

high yields. Although additional energy can be stored as body lipids, this is not so universally true for protein that is undegraded in the rumen and some minerals. Both are required mainly by high-yielding cows. Thus a supplement of undegraded protein and a mineral/vitamin mix might be fed at a low rate, perhaps no more than 1 kg per cow, to high-yielding cows, particularly those with a high level of body condition that can use lipid catabolysis to provide the energy requirements for high milk yields. This could be fed together with a coarse mix of forage and by-products, such as ensiled grains from the brewing industry, from a forage wagon.

The allocation of concentrates to cows that are of different genetic potential should not be confused with cows that are high or low yielding because of their stage of lactation. If all the cows are offered the same concentrate allocation (a flat rate), the reduced allocation in early lactation, relative to cows fed according to their milk yield, will be partly compensated for by increased silage intake. Feeding according to yield probably overfeeds concentrates in early lactation. This is because, firstly, fewer nutrients are required to produce each litre of milk in early lactation as it contains less fat and protein at this stage, and secondly, in early lactation, cows can readily use body tissue that they have stored up during the dry period. Feeding too much concentrate in early lactation can lead to inadequate fibre intakes, low milk fat concentrations, acidosis and lameness. Thus feeding to yield is less suitable for high-concentrate feeding systems, because the cow consumes insufficient fibre. However, a high concentrate:forage ratio can facilitate involution of the digestive tract postpartum, enabling the cow to achieve peak DM intake at an earlier date than a cow fed a high forage ration. This, coupled with the high energy density of the ration, leads to a reversal of the negative energy balance at an early stage of the lactation and a greater chance of the cow becoming pregnant again. Thus a low flat rate of concentrates fed throughout the lactation may underfeed cows in early lactation, compared with feeding to yield, if the forage is of poor quality or restricted in availability.

Whilst acknowledging that flat rate feeding systems may theoretically be better for cows fed a high level of concentrates over the lactation, most dairy farmers would only feed the concentrates at this rate during the winter period. Autumn-calving cows are in mid–late lactation by the time that they are turned out to pasture in the spring, and farmers reduce the concentrate allocation at this time because there is good quality grass available. Spring-calving cows will usually be offered a reduced concentrate ration when they are housed in the autumn because they are in mid–late lactation. Summer-calving cows are likely to require the most concentrate supplement overall, as the grass in late summer is not sufficient to support high yields in early lactation and they will then lactate through the winter on a silage/concentrate mixture. However, in most countries, the milk price is increased in periods when there is little fresh feed available for the cows, depending on the ease of providing conserved food. For example, in sub-Saharan Africa, the milk price typically doubles during the winter dry season, whereas in Britain, a slightly greater milk price is offered in late summer because of the shortage of grazed grass at this time and the reluctance of farmers to start providing their feed conserved for the winter.

Feeding all the cows their concentrates at a flat rate through the winter period has the advantage of simplicity, and it can be fed on top of, or mixed in with their forage, rather than through individual feeders in the parlour or the cow housing. If individual feeders are used for feeding concentrates according to the cow's yield, a careful check should be regularly made that the concentrates delivered by the feeder are the same as the quantities programmed by the farmer, and that once in the feeder the concentrates are consumed by the cow that programmed their delivery, not another cow that enters afterwards. The calibration of the feeders needs to be checked regularly, and in the parlour, some cows, particularly first lactation heifers, cannot consume more than 3–4 kg in the 5 min or so that they spend being milked. Out-of-parlour feeders are often not designed well enough to prevent the cow that triggers release of the concentrate feeds being forcibly ejected from the feeder by another cow, who wants an additional ration to her own. Cows can be fed approximately according to their yield by grouping them into high-, medium- and low-yielding groups and a non-lactating group, which are fed a progressively reduced level of concentrates. As the lactation advances, cows will be moved from the first group to the last, producing a feeding pattern where the concentrate feeding rate is stepped down on two or three occasions. Disadvantages of this 'stepped' feeding system are that the regular movement of cows between groups upsets their social order, and sudden, large changes in concentrate allocation may disrupt rumen function.

The risk of upsetting the rumen digestion with high-concentrate diets and causing low milk fat concentrations or at worst acidosis has led to several developments in concentrate feeding. First, if high levels of concentrate are fed, it is better if they are based on digestible fibre, such as from beets, rather than starch from cereals. The starch in compound pellets is exposed to rapid degradation by the rumen bacteria. Secondly, the concentrates may be fed at a low level several times during the day, rather than just twice during visits to the parlour. This will help to both reduce those bursts of acid production by the rumen bacteria that reduce the efficiency of rumen function and also improve the efficiency of capture of the ammonia by ensuring a more even degradation of the concentrate feed over time. Out-of-parlour concentrate feeders can be programmed to offer each cow her day's concentrate feed in several doses, usually four doses which are available at 6-h intervals. It is clear that cows in such a system know quite accurately when the 6 h is up and it is time for a new allocation, as they can be seen waiting at the feeder. This method of feeding concentrates could be linked to an automatic milking machine to encourage the cows to be milked more frequently.

Complete diets

A complete diet is a mixture of feeds that provides the sole source of nutrients for domestic animals. In cattle feeding it refers to a mixture of forage, usually silage, and concentrates that is made up in a mixer wagon. The wagon costs considerably more than a forage wagon because it has a means of mixing the diet before feeding and often weighing it too. A complete diet can, therefore, be made in the mixer wagon, by loading concentrate supplements or other by-products after putting in the main ingredient, the silage. Mixer wagons are large boxes with a capacity of about 3 t that have an internal mixing system, a delivery chute and usually integral load cells so that the weight of feed that is delivered from the box can be monitored. The mixing system is either large internal agitators (paddles) that mix the diet well in a vertical plane but not horizontally (this can be a problem when small quantities of a mineral/vitamin supplement are added to the mix), or opposing screw augers that move the feed backwards and forwards. Screw augers are likely to compress a wet feed, which then becomes blocked at one end of the wagon. Paddles tease out the feed and are better for handling wet foods. With paddles, it is important to ensure that small amounts of high-cost ingredients are distributed along the length of the wagon. Cows should not be able to select individual feed items from a complete diet, but inevitably some larger food ingredients, such as potatoes, can be preferentially consumed. If there is variation in the consistency of the complete diet, it may be necessary to formulate a premix that includes some of the ingredients that are added in small quantities. If a farmer is feeding his own cereals, a roller mill will be required as cattle cannot utilize whole grains effectively. As well as a mixer wagon, a cattle farmer making complete diets will require good food handling and storage facilities. Food can be conveyed by auger into above ground hoppers, which can then deliver into the mixer wagon by gravity, or it can be stored on concrete in covered yards or, preferably, secure buildings, but pest prevention will be necessary.

Forage boxes are an alternative to mixer wagons. They are large boxes with a similar capacity to mixer wagons, but they only have a movable chain and slat floor internally and a delivery chute. Essentially no mixing takes place in the box. Feeds can be layered into a forage box and some mixing then takes place as the feed is delivered, but selection of individual ingredients by the cows is likely. Because mixer wagons are expensive, some farmers make a simple diet with forage and some by-products in a forage box and supplement it with high-quality supplements rationed on an individual basis to cows that require them, as described above. Machine maintenance is greater for mixer wagons than forage boxes, and they use considerably more power (typically 15, 25 and 35 kW t^{-1} DM are required for a forage box, augered mixer wagon and agitated mixer wagon, respectively).

Complete diets allow cows to be milked without feeding concentrates, reducing the amount of dust in the parlour. Cows that are fed in the parlour are excited by the food reward that they get. This may be a small price to pay when they are at pasture and often difficult to collect if nothing is fed in the parlour. For some high-yielding cows the relief of udder pressure is reward enough, and these usually come in ahead of low-yielding cows.

The main advantage of complete diets is that inexpensive by-product feeds can be incorporated into a mix, and the low palatability of, for example, citrus fruit products can be masked by the strong taste of silage. Concentrate costs may be reduced by about 10% by using 'straight' ingredients, rather than a compound pellet. If good storage facilities are available, advantage can be taken of low feed prices at certain periods of the year, in particular during summer when there is good quality grass available. The benefit that this can bring to a farm must be carefully evaluated, as farmers may be enticed by potential savings in feed costs but not able to capitalize on these if the transport costs are high. They must be able to provide the skilled management and careful rationing required for feeding complete diets to cattle. Typically, it takes about 1 min day^{-1} per cow to feed a complete diet, which is a significant commitment compared with the time normally spent in other activities (milking 2 min per cow, cow movement 0.3 min, parlour cleaning 0.5 min per cow, manure removal 0.7 min per cow, cubicle littering 0.4 min per cow). It takes longer to mix a complete diet if long straw is included. In addition to the time required to mix the diet, the time and expertise must be available to formulate the rations and ensure that the cows receive the optimum diet to meet their nutritional requirements. Feed intakes tend to be high with complete diets and there is a danger of cows getting over-conditioned in this system of feeding. This can lead to fatty liver symptoms and the risk of associated diseases, particularly reproductive failure.

Feed waste can be much greater than normal with complete diets, for example loss of feed to vermin, especially rodents and birds, either in storage or after feeding in the cattle building, particularly starlings, or feed particles being blown around the farmyard when the feed is delivered from a high level storage bin to the mixer wagon in exposed sites. Farms with a high level of infestation by starlings or pigeons may find that cows' backs become dirty, as a result of droppings from roosting birds in the rafters. Buildings can be proofed against bird entry with plastic webbing, but in exposed sites this is not very durable. Feed will keep for several days if the temperature is less than about 10°C, but at high temperatures, mixing needs to be done every day otherwise moulds and yeasts will rapidly form. In a British winter, cows can be fed three times a week (thus avoiding weekend feeding), without the risk of the complete diet spoiling, provided that uneaten feed is regularly returned to the cows. This reduces the aggression between cows, which is associated with feeding times.

In most circumstances nutrients are not utilized more efficiently in a complete diet than when fed separately. However, if high levels of concentrated ingredients are fed, mixing the concentrates with forage that is digested more slowly will reduce the risk of ruminal acidosis. Typical diet formulations for early-, mid- and late-lactation/dry cows are given in Table 2.12.

Conclusions

Feeding systems are important because feed comprises a large proportion of the costs of keeping cattle, typically about 70%, and also because the feeding regime influences product quality, most notably that of milk from the dairy cow. The feeding system also influences the welfare of cattle, especially that of the young calf removed from its mother at an early age. Providing a suitable

Table 2.12. Diet formulations for early-, mid- and late-lactation cows.

	Lactation stage		
	Early	Mid	Late/dry
Yield level (kg day^{-1} per cow)	30–40	20	10
Forage DM as proportion of total DM	0.3	0.5	0.7
Energy density (MJ kg^{-1} DM)	12	11	10
Crude protein (g kg^{-1} DM)	17	14	12
Modified acid-detergent fibre (g kg^{-1} DM)	16	25	30
Calcium (g kg^{-1} DM)	8	6	5
Phosphorus (g kg^{-1} DM)	4.5	3.5	3.0
Magnesium (g kg^{-1} DM)	1.8	1.5	1.5
Sodium (g kg^{-1} DM)	1.8	1.5	1.5

feeding system will prevent ill health and stress, and present the calf with the necessary supply of nutrients to enable it to survive to adulthood.

References

Alderman, G. and Cottrill, B.R. (1993) *Energy and Protein Requirements of Ruminants – An advisory manual prepared by the AFRC Technical Committee on Responses to Nutrients.* CAB International, Wallingford.

Berge, P. (1991) Long-term effects of feeding during calfhood on subsequent performance in beef-cattle (a review). *Livestock Production Science* 28, 223–234

Campling, R.C. (1991) Processing cereal-grains for cattle – a review. *Livestock Production Science* 28, 223–234.

Cheng, K.J., McAllister, T.A., Popp, J.D., Hristov, A.N., Mir, Z. and Shin, H.T. (1998) A review of bloat in feedlot cattle. *Journal of Animal Science* 78, 299–308.

Dellmeier, G.R., Friend, T.H. and Gbur, E.E. (1985) Comparison of four methods of calf confinement. 2. Behaviour. *Journal of Animal Science* 60, 1102–1109.

Fisher, G.E.J., Sabri, M.S. and Roberts, D.J. (1994) Effects of feeding fodder beet and concentrate with different protein contents on dairy cows offered silage *ad libitum. Grass and Forage Science* 49, 34–41.

Frank, B. (1982) Untreated barley straw in dairy cow rations. *Swedish Journal of Agricultural Research* 12, 137–147.

Goering, H.K., Gordon, C.H., Wrenn, T.R., Bitman, J, King, R.L. and Douglas, F.W. (1976) Effect of feeding protected sunflower oil on yield, composition, flavour and oxidative stability of milk. *Journal of Dairy Science* 59, 1416.

Jackson, M.G. (1977) The alkali treatment of straw. *Animal Feed Science and Technology* 2, 105–130.

Kellaway, R.C., Grant, T. and Chudleigh, J.W. (1973) The roughage requirement of the early-weaned calf. *Animal Production* 16, 195–203.

Leaver, J.D. and Fraser, D. (1987) A system study of high and low concentrate inputs for dairy cows: physical and financial performance over 4 years. *Research and Development in Agriculture* 4, 171–178.

Owens, F.N., Scrist, D.S., Hill, W.J. and Gill, D.R. (1997) The effect of grain and grain processing on performance of feedlot cattle: a review. *Journal of Animal Science* 75, 868–879.

Phillips, C.J.C. (1993) *Cattle Behaviour.* Farming Press, Ipswich.

Phillips, C.J.C., Youssef, M.Y.I., Chiy, P.C. and Arney, D.R. (1999) Sodium chloride supplements increase the salt appetite and reduce sterotypies in confined cattle. *Animal Science* 68, 741–748.

Pounden, W.D. and Hibbs, J.W. (1948) The influence of the ratio of grain to hay in the ration of dairy calves on certain rumen micro-organisms. *Journal of Dairy Science* 31, 147–156.

Preston, T.R. and Willis, M.B. (1970) *Intensive Beef Production.* Pergamon Press, Oxford.

Reinhardt, V. and Reinhardt, A. (1981) Cohesive relationships in a cattle herd *(Bos indicus). Behaviour* 77, 121–151.

Roy, J.H.B. (1990) *The Calf. Volume 1, Management of Health,* 5th edn. Butterworths, London.

Rutgers, L.J.E. and Grommers, F.J. (1988) Intersuckling in cattle – a review. *Tidjschrift voor Diergeneeskunde* 113, 418–430.

SandovalCastro, C.A., Anderson, S. and Leaver, J.D. (1999) Influence of milking and restricted suckling regimes on milk production and calf growth in temperate and tropical environments. *Animal Science* 69, 287–296.

Thomas, D.B. and Hinks, C.E. (1982) The effect of changing the physical form of roughage on the performance of the early-weaned calf. *Animal Production* 35, 375–384.

Vérité, R and Journet, M. (1970) Effect of the water content of grass and dehydration at low temperature upon its feeding value for dairy cows. *Annales de Zootechnie* 19, 109–128.

Webster, A.J.F. (1984) *Calf Husbandry, Health and Welfare*. Granada Publishing, London.

Wright, I.A., Rhind, S.M., Whyte, T.K. and Smith, A.J. (1992) Effects of body condition at calving and feeding level after calving on LH profiles and the duration of the postpartum anestrus period in beef cows. *Animal Production* 55, 41–46.

Further Reading

Buchanan-Smith, J.G. and Fox, D.G. (2000) Feeding systems for beef cattle. In: Theodorou, M.K. and France, J. (eds) *Feeding Systems and Feed Evaluation Models*. CAB International, Wallingford, pp. 129–154.

Chamberlain, A.T. and Wilkinson, J.M. (1993) *Feeding the Dairy Cow*. Chalcombe Publications, Lincoln.

Owen, J.B. (1991) *Cattle Feeding*. Farming Press, Ipswich.

Tamminga, S. and Hof, G. (2000) Feeding systems for dairy cows. In: Theodorou, M.K. and France, J. (eds) *Feeding Systems and Feed Evaluation Models*. CAB International, Wallingford, pp. 109 127.

Webster, A.J.F. (1993) *Understanding the Dairy Cow*. Blackwell Scientific, Oxford.

Nutrient Requirements and Metabolic Diseases

3

Introduction

Determining the nutrient requirements of cattle to enable them to be effectively rationed has been a focus of attention for nutrition scientists for nearly 200 years. As genetic selection has increased the potential productivity of cattle, particularly dairy cows, so their nutrient intake has to increase. If this cannot meet the high requirements for production, dairy cows may lose a lot of weight. Too great a difference between requirements and nutrient supply leads to either metabolic breakdown or a reduction in production levels. The greatest nutrient deficit occurs during the first few weeks of lactation in the dairy cow, mainly because breeders have selected cows that increase their milk production rapidly over the first few weeks. This often overloads the ability of the digestive system to supply adequate nutrients, resulting in metabolic breakdown. In future, breeders are likely to select cattle with more persistent lactations, rather than a rapid increase to peak lactation. This could lengthen the lactation and prevent the production of unwanted calves where the demand for milk is large relative to the demand for beef. In the long term, it may be possible to regulate the apoptosis of mammary gland cells, perhaps leading to cows that are permanently lactating after their first calf.

Dairy cow managers must provide foods that enable the cow to increase intake as rapidly as possible after calving. These should be highly digestible feeds and the manager should also ensure that the way in which they are offered is conducive to high intakes. Complete diets, where the animal's entire diet is offered in one well-mixed feed, are one of the best ways to achieve this. For very high-yielding cows, self-feed silage and tight grazing systems are probably best avoided.

Diets can now be accurately constructed to enable both male and female cattle to grow or lactate at various levels, largely as a result of the extensive research effort of the latter part of the 20th century. Efficient rationing is particularly important in countries where milk output is restricted by quota, but can also be vital in countries where there is strong competition between milk producers for market share or where large units prevail, creating the economics of scale that can allow the purchase of sophisticated equipment for feeding and rationing the animals. In such situations, accurate rationing will enable farmers to adjust milk output and quality according to the market conditions. In developing countries, the shortage of high quality feed resources for cattle feeding makes effective rationing of major importance, but the research effort into managing cattle nutrition in tropical conditions in particular has not yet been nearly as extensive as in the temperate zones.

Least cost rationing, where diet ingredients are compared in terms of the nutrients that they can supply and their cost, can now be rapidly formulated with the aid of a computer, which solves a series of quadratic equations to arrive at the optimum diet. Such methods are already essential in devising rations for pigs and poultry, but are necessarily confined to feeds that can be purchased, or attributed a cost, i.e. mainly conserved feeds, rather than grazed grass. A cost can be introduced for forage, which is saleable, but if it is home produced, the opportunity cost of the land must be a prime consideration. Maximum and minimum inclusion rates can be included for certain ingredients to take into account particular constraints, such as the bulk density of the feed, its palatability or subtle aspects of its digestion or utilization, such as the presence of antinutritive agents, which may not be accounted for in the rationing system. In large, sophisticated dairy farming systems, complete diets are often mixed centrally and transported to farms within a driving distance of 2–3 h. The mixing factory can take advantage of low cost feeds, bulk purchase of ingredients and the best nutritional expertise to formulate rations for the farms.

In cattle farming systems where the animals spend the summer at pasture and are given conserved feeds in winter, it is essential that forage stocks are assessed well before the end of the summer. Silage volume can be calculated from a knowledge of clamp, tower or bale dimensions and converted to a mass using known densities of silages of different dry matter (DM) concentrations. The DM content should be estimated by taking cores at regular intervals through the silage. As silage DM content varies considerably at different points in the clamp, it is not sufficient to sample silage at the edge of the clamp. Tabulated values of silage density can then be consulted. For example, grass silage at 220 g kg^{-1} DM contains approximately 700 kg fresh silage and 155 kg silage DM m^{-3}. At 300 g kg^{-1} DM, grass silage contains approximately 650 kg fresh silage but 195 kg silage DM m^{-3}. Maize silage tends to have lower densities than grass silage. Silage in towers has a high density because of the greater height of silage squeezing air out very effectively. Conversely, big bales of silage have a low density because of lack of air removal. The DM density is most important for rationing purposes, but the wet matter density is also necessary for weighing out rations.

In addition to a knowledge of silage DM and fresh matter consumption by the cattle, the silage should be chemically analysed to obtain information concerning its nutrient value for feeding to cattle. As with determinations of DM concentration, it is most important to use the proper sampling techniques. The silage analysis should determine or estimate the concentrations of DM, crude protein, digestible organic matter (DOM), metabolizable energy (ME), pH, ammonia-nitrogen as a proportion of total nitrogen, and ash. To determine the contribution to metabolizable protein (MP), the fermentable ME, effective rumen-degradable protein and digestible undegradable protein should also be predicted.

From a knowledge of silage quality and quantity, the farmer can approximately predict the performance of his cattle, given a certain level of concentrate supplement, or can estimate how much concentrate supplement to feed in order to achieve a certain production level. This has been made possible by recent research that has focused on changing the old systems of estimated cattle nutrient 'requirements' into prediction systems where responses are determined to additional nutrients. Cattle can only be said to have a 'requirement' for a certain nutrient if they respond up to a certain point and then no more, whereupon another nutrient becomes limiting. This is analogous to the control of photosynthetic rate in plants by temperature, CO_2 concentration or irradiance. In this case only one resource will limit output at any time, which can be increased in a stepwise fashion by providing more of each of the resources in turn above its limiting threshold. In animals, this response pattern to a series of different nutrients rarely occurs, and it is an oversimplification of plant responses. Because animals can store most of the major nutrients for long periods of time, in preparation for future restrictions, the instantaneous response is often quite different from the long-term response. Also, nutrients interact with each other; for example, minerals that are similar in their electronic configuration often compete with each other for absorption or adsorption sites. In addition, nutrients may not always be used for the same purpose, as for example protein, which can be used for either the nitrogen or energy demands of the animal.

Rationing grazing cattle presents particular problems, as it is difficult to know how much the animals are eating, even in research. Sampling grazed herbage is also more difficult than conserved forage as cattle are more selective than when they are offered conserved forages, and it is difficult to know precisely what they are eating.

Responses will partly be determined by the animal's genotype, which influences the partitioning of nutrients and feed intake, but this cannot yet be reliably included in rationing systems. Most attention has focused recently on improving protein response systems, given the complexity of protein metabolism in the ruminant. However, energy response systems are at least as important in determining the output and profitability of cattle production units, but the rationing systems were developed in advance of the protein-rationing systems.

The British Energy Response System

Energy is usually the first limiting nutrient in a diet for cattle. Because the cattle respond to increases in energy intake even when a high energy diet is fed, it is

Table 3.1. The distribution of energy in a feed for dairy cows and the determination of digestible, metabolizable and net energy concentrations of the feed.

Dry matter intake (kg day^{-1})	18
Gross energy intake (MJ day^{-1})	350
Faecal energy (MJ day^{-1})	135
Urinary energy (MJ day^{-1})	12
Methane energy (MJ day^{-1})	24
Heat increment (MJ day^{-1})	70

not possible to determine an energy 'requirement', nor even an allowance, that is not dependent on the production level. However, dose–response relationships allow production to be predicted when energy intake is known, or energy requirements to be determined for a specific production level. For example, the milk yield can be predicted for the cows in a dairy herd, either as a group or individually, using the intended diet and the appetite of the cows to predict energy intake. After the energy requirements for maintenance, weight changes and pregnancy (if applicable) have been deducted, the residual energy can be assumed to be that available for milk production. As long as the composition is known or can be anticipated, and the energy requirement per litre of production estimated, yield can be predicted. If the actual yield is then monitored regularly, any deviations from the predicted yield can be investigated. On an individual basis, such deviation may indicate ill health, and on a group basis the energy rationing system can be used to predict how feed energy supply should be altered to bring the actual yields back in line with predicted yields. This is especially important for quota management.

Comparative analysis of agricultural feeds to promote the fattening of cattle was instigated by Kellner at the end of the last century. He produced 'starch equivalent' values, which related the fattening ability of different feeds to that of pure starch. The subsequent values obtained were used for much of the 20th century to ration cattle. In 1975 the British Ministry of Agriculture published energy allowances for cattle based on ME, a system which is essentially still in use today (Ministry of Agriculture, Fisheries and Food, 1975). The ME content of a feed was defined as the energy content remaining after faecal, urine and methane energy have been subtracted (Table 3.1).

$$\text{Digestible energy content} = \frac{350 - 135}{18} = 11.9 \text{ MJ kg}^{-1} \text{ DM} \qquad (3.1)$$

$$\text{Metabolizable energy content} = \frac{350 - (135 + 12 + 24)}{18} = 9.9 \text{ MJ kg}^{-1} \text{ DM} \qquad (3.2)$$

$$\text{Net energy content} = \frac{350 - (135 + 12 + 24 + 70)}{18} = 6.0 \text{ MJ kg}^{-1} \text{ DM} \qquad (3.3)$$

The net energy is that which is used to maintain the animal, for lactation, body growth and pregnancy, however, the ME fraction is used because of the difficulty of measuring the heat increment. The main energy source in cattle feed is structural carbohydrate, which has a greater heat increment than other energy sources. Protein and fat have greater energy concentrations than structural carbohydrate, but the former is used largely to supply the protein requirements, and the latter cannot be included in high concentrations in the diet of ruminants because of adverse effects on rumen bacteria.

The cornerstone of energy rationing is the assumption that energy intake, from forages, concentrate and body tissue mobilization, is equivalent to energy output, in milk, maintenance, pregnancy and live weight gain. This has to be approximately true overall, since energy can neither be lost nor created. However, the factorial addition of energy contributions from different sources is not absolutely accurate, since as energy intake increases the utilization of the energy becomes less complete, partly because more rapid passage through the gastrointestinal tract reduces the feed digestibility. Thus, one additional kilogram of concentrate will have a lower energy value if it is added to the diet of a high-producing cow that is already consuming an energy-dense diet, than if it is added to the diet of a cow fed at maintenance. Another problem with the factorial energy rationing system is that it is not possible to predict the fate of additional energy consumption. It may contribute to increased milk production, but it may also contribute to lipogenesis. Much depends on the source of the energy, and in particular whether it provides lipogenic or glucogenic precursors, and the production level of the cattle. However, the ME system has been used successfully for rationing cattle now since 1975, and is believed to be reasonably accurate. By contrast, protein-rationing systems have fundamentally changed since this time.

The factorial estimation of energy output is determined as follows: maintenance energy requirements are based on energy utilization during fasting, adjusted for the weight of the animal and an activity increment and divided by the efficiency of ME utilization for maintenance. The fasting energy utilization is assumed to be greater for bulls than other cattle, in part because of their reduced level of subcutaneous fat. The activity increment can be increased for grazing compared with housed cattle. The energy costs of different activities, getting up, walking, lying down, etc., have been determined and the maintenance requirements can be accurately calculated, if necessary. Fasting energy requirements can be predicted as follows:

$$\text{Energy utilized during fasting (MJ day}^{-1}) = C1\,[0.53(W/1.08)^{0.67}] \qquad (3.4)$$

where $C1$ is a constant of 1.15 for bulls and 1.0 for other cattle, and W is weight in kg.

The activity increment is also based on the animal's weight, and is estimated as follows:

$$\text{Activity increment (MJ day}^{-1}) = 0.0095 \times \text{weight (kg) for lactating cows, and}$$

$$0.0071 \times \text{weight (kg) for other cattle} \qquad (3.5)$$

The efficiency of energy utilization for maintenance is dependent on the metabolizability of the diet, i.e. the concentration of ME, and is typically about 70%.

The energy requirements for milk production are well established for milk of different compositions. They can be calculated by multiplying the milk yield by the energy value of the milk, which can be determined from the milk composition as follows:

$$\text{Energy value (MJ kg}^{-1}) = 0.038\ F + 0.022\ P + 0.020\ L - 0.11 \tag{3.6}$$

where F, P and L are the fat, crude protein and lactose concentrations in milk in g kg^{-1}.

For lactating cows, the energy required for weight change will depend on the composition of the gain. A late lactation cow that is replenishing body fat stores will require more feed energy per kilogram live weight gain than a heifer that is still growing during her first lactation, since the cow will be laying down fat tissue whereas the heifer will be depositing other tissues, such as muscle, which have a lower energy content. Unfortunately, it is not possible at present to account for the variation in the energy requirements for weight change in lactating cows, which can be considerable, and a common value of 19 MJ kg^{-1} live weight change should be adopted. If body tissue is mobilized, it will be used with an efficiency of approximately 84%, thus contributing $19 \times 0.84 = 16$ MJ day^{-1} of feed energy equivalent.

The energy required for the growth of beef cattle is also dependent on the composition of the gain, and can be predicted from the rate of gain, with a correction for the class and breed of cattle (Table 3.2).

Bulls produce leaner growth than steers at a particular weight, which in turn produce leaner growth than heifers, and lean tissue deposition requires less feed energy than fat tissue. In this analysis, Aberdeen Angus and North Devon are classified as early maturing, Hereford, Lincoln Red and Sussex as medium maturing and Charolais, Limousin, Simmental, Friesian and South Devon as late maturing. The equation for predicting the energy value of live weight gain is as follows:

$$\text{Energy value (MJ kg}^{-1}) = \frac{C1(4.1 + 0.0332\ W - 0.000009\ W^2)}{(1 - C2'0.148\Delta W} \tag{3.7}$$

where $C1$ is the correction factor given in Table 3.2, and $C2$ is 1 where the plane of nutrition is sufficient to provide for at least maintenance, otherwise it is 0. W is the weight, in kg.

The energy requirement for gestation is a function of the gestation stage and the potential for growth of the fetus, as determined by its breed. It can be calculated from the following equation:

$$\log_{10} \text{energy requirement} = 151.665 - 151.64\ e^{-0.0000576t} \tag{3.8}$$

Table 3.2. Correction factors for the energy content of weight gain by beef cattle.

Maturity classification	Bulls	Steers	Heifers
Early	1.00	1.15	1.30
Medium	0.85	1.00	1.15
Late	0.70	0.85	1.00

where t is days from conception. Requirements can be adapted for different breeds of calf.

The summated energy requirements can then be compared with energy supply, which is determined from feed intake and the energy concentration of the feed. If the diet is hypothetical, and the farmer is predicting how much his cattle will consume, this must be tested against equations that estimate how much they should be capable of eating, given the various cow and feed factors involved. The simplest equations use the milk yield of the cows and their weight to predict intake, but more complicated equations exist that introduce feed factors. The following equation gives one of the best estimates of grass silage DM intake under British conditions:

$$\text{Silage DM intake} = -3.74 - 0.387C + 1.486\ (F + P) + \\ \text{(kg day}^{-1}) \quad\quad 0.0066\,W + 0.0136\ \text{DOMD} \quad\quad\quad (3.9)$$

where C is concentrate DM intake in kg day^{-1}, $F + P$ is the daily yield of fat and protein in milk in kg day^{-1}, W is the cow's weight in kg and DOMD is the digestible organic matter in the DM in g kg^{-1} DM.

If the proposed ration is unlikely to be able to be consumed by the cows then the energy concentration must be increased to achieve the required energy intake or the productivity of the cows will decline. The ration must be offered in such a way as to maximize DM intake in early lactation.

An example of the use of the British energy response system by a veterinarian

A veterinarian is called to a dairy farm in December where the farmer has recently employed a new herdsman and has a problem with rebreeding his autumn-calving herd. Many of the cows are not pregnant and the farmer suspects a large proportion of his cows are not cycling properly for one of two reasons:

1. Some cows are believed to have ovarian cysts – the veterinarian inspects a number of cows and finds actually there are few cows with this condition.
2. The cows' ration contains inadequate energy and the prolonged negative energy balance of the cows has reduced conception rates – a view proposed by the farmer's feed merchant.

On finding that the first suspected cause of cow infertility was not likely, the veterinarian is asked to comment on the feed merchant's claim that the ration for the cows had inadequate energy contents. She decides to check the energy balance.

The mean weight of the Friesian cows is estimated to be 650 kg. They are producing a mean milk yield of 33 kg, with a butterfat content of 38 g kg^{-1}, protein content of 32 g kg^{-1} and a lactose content of 50 g kg^{-1}.

The foods available to the farmer are silage containing 200 g DM kg^{-1} and 11 MJ ME kg^{-1} DM (730 g DOM kg^{-1} DM), rolled barley containing 850 g DM kg^{-1} DM and 13.7 MJ ME kg^{-1} DM and dairy compound containing 860 g DM kg^{-1} and 12.5 MJ kg^{-1} DM. The farmer has estimated that he is feeding 50 kg silage, 4 kg barley and 6 kg compound feed per cow per day.

The questions that the veterinarian must answer are:

1. What are the cows' ME requirements? (These are early lactation cows so she assumes that they will be losing about 0.5 kg day^{-1} body weight at 16 MJ kg^{-1} of feed equivalent; if cows lose weight at more than 0.5 kg day^{-1} conception rates are likely to be adversely affected.)

2. Is the ration within appetite limits? If it is not, the farmer's estimated intake per cow may be too high.

3. What is the ME provided by the ration and does it match energy requirements?

The veterinarian's findings:

1. The ME content of the milk is 5.1 MJ l^{-1} and milk ME output is therefore 168 MJ day^{-1} (Equation 3.6). The ME requirement for maintenance is 61 MJ (Equations 3.4 and 3.5). The total ME requirement can be reduced by 8 MJ day^{-1}, because the cows are estimated to be losing 0.5 kg day^{-1}. Therefore the total ME requirement is 221 MJ day^{-1}.

2. The potential intake of silage DM is 10 kg day^{-1} (Equation 3.9), so the farmers estimate of the cows' intake may be correct.

3. The daily ME intake is:
   silage, 10 kg DM = 110 MJ
   barley, 3.4 kg DM = 46.6 MJ
   compound, 5.2 kg DM = 64.5 MJ
   Total, 18.6 kg DM = 221 MJ.

Sustaining this level of body tissue loss may reduce conception rates if the cows' body condition is already at a low level, but the ration that the farmer is feeding to his cows appears feasible. The next step would be to score the cows' body condition. If many are below 2, on the five point UK scoring system, then she should recommend increasing the concentrate part of the diet to increase energy intake.

If they are not, she should investigate the oestrus detection rate (number of cows expected to be in oestrus divided by number of cows observed in oestrus). If it is less than 60%, poor oestrus detection by the new herdsman is likely to be the reason for the problem.

The US Carbohydrate Fractionation System

The Cornell group in the United States have developed a system of fractionating carbohydrates consumed by cattle that allow the different rates of degradation of the carbohydrate fractions to be predicted and incorporated into feeding standards. The total carbohydrate content of a feed is divided into neutral detergent soluble and insoluble components. The former include fraction A (sugars and organic acids), which is highly degradable, and fraction B1 (starch and pectin), which is of intermediate solubility. Fraction B2 is available plant cell wall and is slowly degradable. This fraction is particularly large in forages. Fraction C is unavailable plant cell wall, essentially lignin, which may reduce the availability of some essential minerals, such as iron, by providing organic ligands that sequester the mineral ions and prevent them from being digested.

Nitrogen Response Systems

Effective management of nitrogen inputs to the dairy cow is important because of the high cost of proteinaceous feeds, the relationship between protein intake and several metabolic diseases, notably reproductive disorders (Butler, 1998) and lameness, and the cost to the environment of high nitrogen emissions from cattle, particularly dairy cows.

For much of the 20th century, protein rationing was based on estimates of digestible crude protein requirements, obtained from empirical data of cattle producing different quantities of milk or growing at different rates when fed varying levels of digestible crude protein. However, this was unsatisfactory for the high-yielding dairy cow in particular, because the extent to which the consumed protein is degraded in the rumen partly determines the amino acid supply to the animal. If feed protein is extensively degraded, the capacity of the rumen microbes to utilize all the degraded nitrogenous compounds is exceeded, and the surplus is absorbed as ammonia and converted by the liver to urea for recycling into the gastrointestinal tract or excretion. However, if some of the protein escapes degradation in the rumen, then providing that the requirements of the rumen microbes are met, the undegraded proteins would pass through to the duodenum for absorption. These then contribute directly to the animal's amino acid requirements, in addition to the microbial protein which is made available to the animal when the microbes are digested by enzymes and the nitrogen breakdown products absorbed.

Nitrogen response systems available for use on farms improved considerably when, in 1980, the British Agricultural Research Council (ARC, 1980) published a system that recognized the importance of independently quantifying the contributions to amino acid requirements from microbial and directly absorbed (rumen-undegraded) sources. Since that time the system has been refined considerably and many other countries have published similar systems (Table 3.3).

These systems now can be used in the field with reasonable success. The essential features of the UK protein-rationing system were that microbial protein production could be predicted from the energy supply to the rumen, and that the requirement for directly absorbed protein could be predicted by subtracting microbial protein supply from the total protein requirements. This was predicted factorially from the requirements for maintenance, body tissue growth and milk protein production. Thus if the whole tract protein digestibility is known, enabling the faecal nitrogen loss to be calculated, and the degradability of the dietary nitrogenous compounds in the rumen,

Table 3.3. The latest systems of protein rationing developed for cattle in specified countries.

Country	Protein-rationing system	Reference
UK	Metabolizable protein (MP)	AFRC (1992); Alderman and Cottrill (1993)
Australia	Apparently Digested Protein Leaving the Stomach (ADPLS)	AUS (1990)
France	Protein Digested in the Intestine (PDI)	INRA (1989)
Germany	Crude Protein Flow at the Duodenum	Ausschuss für Bedarfsnormen (1986)
Holland	Digestible Protein in the Intestine (DVE)	Central Veevoederbureau (1991)
Norway	Amino acids truly absorbed in the small intestine and protein balance in the rumen (AAT-PBV)	Madsen (1985)
Switzerland	Absorbable Protein in the Intestine (API)	Landis (1992)
USA	Cornell Net Carbohydrate and Protein System (CNCPS)	Sniffen *et al.* (1992)

enabling the nitrogen utilized by the rumen bacteria or recycled as urea to be predicted, the supply of digestible undegraded nitrogen can be determined.

More recently, the introduction of the MP system in Britain and similar systems in other countries have brought a range of improvements in protein rationing. Among these are the recognition that only *fermentable* energy will contribute to microbial growth in the rumen. Metabolizable energy from feeds that have already been fermented will be partly composed of the acid end-products of fermentation, estimated to be about 10% of the total ME for silage and 5% for distillery by-products, which will not provide energy substrates for microbial growth. However, different forms of fermentable metabolizable energy (FME) are used with different efficiencies for microbial maintenance and growth, which is not considered by the British system.

The British system does recognize that microbial growth may be limited by rumen-degradable protein supply, not energy, and a further innovation was the division of rumen-degradable protein into slowly and rapidly degradable nitrogen. Only 80% of the latter is estimated to contribute to the nitrogen requirements for microbial growth, because some of the rapidly degradable protein and even more of the non-protein nitrogen will be absorbed before it can be captured by the rumen microorganisms, particularly if it is consumed rapidly and immediately solubilized, leading to a surge in the ammonia concentration in the rumen. The rapidly degradable protein can be mathematically predicted from the DM disappearance pattern of feeds suspended in the rumen in fine mesh nylon bags *per fistulam*. This has inherent errors, as the bag will allow some unfermented small particles to escape and some particles from outside the bag to enter, it does not allow the feed to be mixed with the other rumen contents, and it may result in selectivity in the rumen microbial population that ferments the food. Despite these sources of inaccuracy, the *in situ* method of estimating protein degradation in the rumen has often correlated well with *in vivo* measurements of the flow of nitrogen fractions from the rumen, but some reservations remain about its use for forages. In future *in vitro* methods are likely to assume more importance, and may be based on the solubility of protein, the incubation of feedstuffs in rumen liquor obtained from an abattoir or near-infrared spectroscopic analysis.

Accepting that the *in situ* method is one of the best currently available, an exponential equation can be used to describe the loss of nitrogen compounds from the bag over a period of 48–70 h (Fig. 3.1):

$$\text{Degradability} = a + b[1 - e^{(-ct)}] \qquad (3.10)$$

where a = water soluble nitrogen, b = insoluble but potentially degradable nitrogen compounds, which are degraded according to first order kinetics with rate constant, c, and c = fractional rate of degradation of nitrogen compounds per hour.

The potentially degradable nitrogen compounds are transformed into actually degraded nitrogen compounds by estimating the retention time of feeds in the rumen.

$$P = a + \{(b \times c)/(c + k)\} \qquad (3.11)$$

where P = the effective degradability of a feed at a rumen outflow rate (k, number per hour) and b and c are as defined above. The retention time is largely a function of the level of feeding of the cattle in relation to their size, and AFRC (1992) recommend the use of the following formula related to the level of feeding, L, which is expressed as a multiple of the ME requirement for maintenance:

$$\text{Rate of outflow (number per hour)} = -0.024 + 0.179\,[1 - e^{(0.278L)}] \qquad (3.12)$$

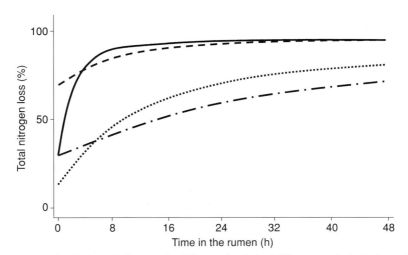

Fig. 3.1.   Degradation of nitrogen in different feeds for different periods in the rumen. Key: —— barley, ···· hay, – – – grass silage, — · — · fishmeal.

High-yielding cows have a high level of feeding and short retention time, whereas mature beef cattle are likely to have a low level of feeding and longer retention time.

The MP system also allows for the fact that some protein that is undegraded in the rumen will not be absorbed at all. This is currently determined chemically as acid-detergent insoluble nitrogen (ADIN). Feeds containing high concentrations of tannins contain ADIN and will have reduced protein digestibility, because of the formation of indigestible tannin–protein complexes in the gastrontestinal tract. Distillery by-products also contain considerable amounts of ADIN if they have undergone prolonged heating during the distilling process. ADIN from distillery by-products appears to be partly digestible, but this is not yet incorporated into the MP system.

If the microbial true protein supply is added to the supply of protein that is undegraded in the rumen, an estimate of the total MP supply is obtained (Fig. 3.2). Alternatively, if the MP supply can be estimated factorially from the N requirements for maintenance and production, endogenous N secretions and urinary N, the requirement for rumen-undegraded protein can be determined by subtracting the microbial protein supply from the MP. The efficiency of utilization of MP for productive purposes is about 85% if the amino acid balance is optimal. It invariably is not, so values of 60–85% are

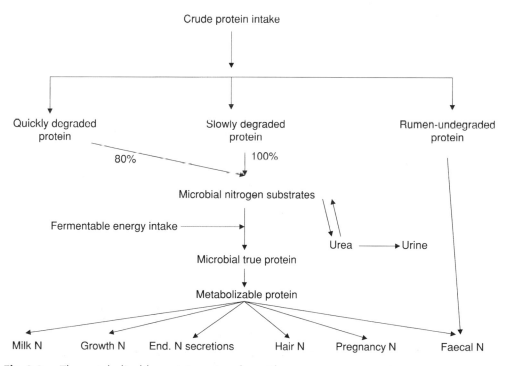

Fig. 3.2. The metabolizable protein system for cattle.

used for the necessary calculations. However, published experiments have suggested that supplementary protein is utilized much less efficiently than this for the production of milk protein, probably because much of the additional protein is used as an energy source. The utilization of protein as energy and nitrogen sources will need to be incorporated into the MP system, and future research should concentrate on modelling the effects of different ratios of lipogenic:glucogenic:proteogenic precursors on the production of milk constituents and body tissue changes in cattle.

Other deficiencies in the current protein-response system that have yet to be satisfactorily resolved are:

1. The increase in feed intake that can occur when protein intake increases. This may also be caused by the use of the additional protein as an energy supply, fuelling increased microbial growth.
2. The energetic cost of converting surplus ammonia to urea in the liver.
3. The optimum amino acid proportions in microbial protein for the different purposes for which it is used.
4. Predicting the rates at which rumen microorganisms degrade protein from different feeds.

Further refinement of the system should enable its value to the dairy farmer in particular to increase. Better definition of nitrogen capture by rumen microbes is already possible and is incorporated into the system devised by Cornell University, USA. Fractionated protein nitrogen is divided into three classes – A (non-protein nitrogen), B (true protein) and C (unavailable insoluble nitrogen). Fraction A is determined as trichloracetic or tungstic acid-soluble nitrogen and contains nitrate, ammonia, amines and free amino acids, which are instantly degraded in the rumen. Fraction B is subdivided into B1 (globulins and some albumins that are soluble in a buffer and are rapidly degraded in the rumen), B2 (most albumins and glutelins that are slowly degraded in the rumen at about 10% h^{-1}, but are all digested in the small intestine) and B3 (prolamins, extensin proteins and heat denatured proteins that have not undergone the Maillard reaction, which are degraded slowly in the rumen at 0.1–1.5% h^{-1} and are 80% digested in the small intestine). Fraction B3 is determined as soluble in acid detergent but not neutral detergent, and B2 is the difference between buffer-soluble protein (B1) and neutral detergent insoluble protein. The system is particularly useful to evaluate high DM silages, where much of the nitrogen is in the A and C fractions and therefore provides little available true protein. The system could also predict the optimum amount of heating to maximize the B3 fraction, which would be of value to high-yielding cows in particular. The B3 fraction can be increased by mild heating but decreased by excessive heating.

In future, it should be possible to predict the MP concentration in a feed by conducting laboratory analyses, in particular the crude protein concentration, the non-protein nitrogen concentration, the solubility of the ingested protein in a buffer solution and the acid-detergent insoluble protein, rather than using *in situ* techniques requiring bags to be inserted into the rumen of fistulated animals. The role of peptides requires further research (Webb *et al.*, 1993). Mechanistic models will become available to predict the MP value, which will incorporate a better prediction of passage rate through the rumen than presently used, using rate-limiting factors such as particle size and density and hydration rate to predict the rate of passage of the solid and liquid fractions of the rumen contents separately.

Mineral and Vitamin Responses

Close relationships between a number of cattle disorders and the supply of certain minerals and vitamins are now well established. Many mineral disorders are locally well known and arise from regional soil deficiencies. These have often been given a variety of local names in different regions. Recently, the reduced reliance on home-grown feeds and greater use of purchased feeds, which may have come from areas that are not deficient in the same minerals and are frequently fortified with minerals and vitamins, has reduced the prevalence of locally recognized disorders. In addition, a better understanding of the aetiology of the diseases has led to effective preventative measures being taken on many farms, particularly mineral supplementation. However, it is now becoming clear that subclinical deficiencies of many minerals and vitamins can reduce cattle performance and impair the immune responses to disease challenges. Such deficiencies are difficult to identify and treat, but the cost:benefit ratio of correctly doing so can be high.

Compared with growing cattle, the dairy cow is particularly at risk of mineral deficiencies, especially in early lactation when her inability to meet the increased demands for lactation may limit production and can cause metabolic disease. Supplementary minerals and vitamins that are absorbed easily and in the correct balance are now available and can be most easily added to proprietary compound feeds or complete diets, although some care is required to ensure adequate mixing in the case of the latter.

A crucial time in relation to the mineral supply for dairy and beef cattle is in spring when they are turned out to pasture after winter housing. The rapid growth of grass reduces the concentration of many minerals, and the rapid passage of herbage through the gastrointestinal tract causes some minerals to bypass the rumen where much of the absorption takes place. At this time of year the effect of mineral fertilizers on herbage composition has also to be carefully considered, as the application of potassium in particular can disturb the balance of other essential minerals in herbage. The minerals that are most likely to be deficient in high-yielding cows are calcium, phosphorus, magnesium and sodium. However, the mineral concentrations in forages vary considerably, depending on where they were grown and what fertilizers were used. Some minerals, e.g. selenium, may be both toxic and deficient to ruminants, depending on the concentrations in feeds, which in turn depends on soil mineral concentrations and availability. The water supply may also contain a significant quantity of minerals and should not be forgotten in calculating requirements.

Vitamins tend to function as catalysts or co-enzyme factors in metabolism and are required in ultra-trace quantities. Most natural feeds, especially grasses, contain adequate supplies of vitamins, or the precursors which allow the vitamins to be synthesized in the rumen, but the storage and preservation of feeds often reduces their vitamin content, making supplementation essential for productive stock.

Major minerals

Calcium

Mammals have elaborate calcium homeostatic mechanisms as a result of the need for its rapid mobilization during lactation. To achieve this they have to accumulate calcium when they are not lactating, which is mainly stored in bones. During lactation the production of parathyroid hormone (PTH) increases in response to low plasma calcium concentrations in order to increase both absorption from the gastrointestinal tract and mobilization from bone. Conversely, when supply exceeds demand the antagonist of PTH, calcitonin, reduces absorption and increases calcium accretion into bone. Vitamin D is also part of the control mechanism, since metabolites of the vitamin that are produced in the liver [25(OH) vitamin D] also enhance calcium mobilization from bone. The PTH also stimulates renal absorption of calcium from the glomerula filtrate. It is only when the deficiency is severe that it promotes the secretion of $1,25(OH)_2$ vitamin D by the kidney, which increases calcium absorption by increasing the production of calcium-binding proteins. Some hypocalcaemic cows have equivalent concentrations of PTH to normal cows, but their kidneys do not respond to the PTH signals.

Calcium homeostatic problems occur mainly in high-yielding dairy cows on the first day of lactation, with an incidence of hypocalcaemia of 4–10% in dairy cows. Still greater economic loss probably occurs because of subclinical hypocalcaemia. At the beginning of lactation the depletion of calcium status occurs suddenly, over about a 10-h period. The large outflow of calcium in milk, together with a depressed appetite around parturition, has the potential to create a calcium imbalance that rapidly leads to paralysis (parturient paresis). Colostrum contains about 2 g Ca l^{-1} and its production at the start of lactation rapidly depletes the body pool of about 12 g of available calcium. In the absence of dietary manipulation of calcium intake, the initial response by the cow to the calcium deficiency is to increase gastrointestinal absorption. It is not until about 10 days later that bone resorption increases (Fig. 3.3). Parturient hypocalcaemia or paresis, which is commonly known as milk fever, may ensue if the imbalance cannot be corrected. Intravenous calcium treatment with 8–11 g Ca should be given to hypocalcaemic cows to keep them alive for long enough for them to activate their calcium homeostatic mechanisms. The best method of prevention is to encourage the cow to begin

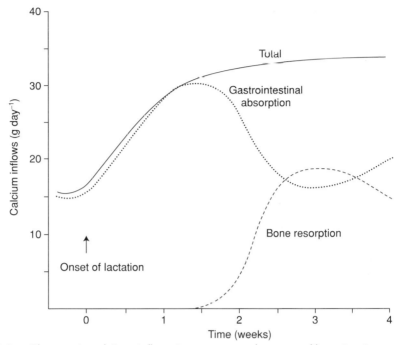

Fig. 3.3. Changes in calcium inflows in response to the onset of lactation in cows fed a diet with normal calcium content.

mobilizing calcium from bone reserves before parturition. If calcium intake can be limited to 50 g day^{-1} or less during the latter part of the non-lactating period, PTH production increases and the mobilization of the substantial bone reserves starts before the onset of lactation. This can also be achieved with large doses of vitamin D before calving. During lactation, calcium intake should be about 3 g kg^{-1} milk.

Calcium absorption is not just determined by homeostatic mechanisms but also by the concentrations of other minerals in the gastrointestinal tract. In particular, very high or low concentrations of phosphorus in feed in relation to calcium restrict its absorption, and a ratio of calcium to phosphorus of 1:1 appears optimal for calcium absorption. A high concentration of magnesium also restricts absorption of calcium. An anionic excess in the diet, in particular a high content of chloride or low content of potassium, may be beneficial if it acidifies the blood, since this enhances the 1,25(OH)$_2$ vitamin D response to hypocalcaemia. Sulphur is not very active ionically, so chloride salts are more appropriate to alter blood pH, but may be unpalatable unless they are mixed into a complete diet. This technique of modifying acid:base balance does not always prevent milk fever, and in several experiments varying the cation:anion balance has failed to alter blood pH, which is normally effectively regulated by homeostatic mechanisms. Perhaps of greater importance than effects on blood pH are direct effects of potassium on the absorption of calcium, via the electrochemical transepithelial gradient. It is already well known that potassium inhibits the absorption of magnesium in the gastrointestinal tract by this mechanism, but it also inhibits calcium absorption in the intestine and its resorption in the renal glomeruli. Ideally for cattle, the potassium concentration in herbage should be less than $10 \, g \, kg^{-1}$, but this is likely to inhibit plant growth. However, if sodium replaces some of the potassium fertilizer, the sodium can assume some of the functions of potassium in the plant and will reduce potassium concentration to the benefit of the ruminant consumer.

Hypocalcaemia often occurs concurrently with hypoglycaemia, particularly in cows calving out at pasture in the autumn, when the herbage has a low DM concentration and limited nutritional value. Feeding forage supplements to cows about to calve will help to reduce the possibility of a calcium imbalance occurring. Hypocalcaemia is also associated with ketosis and mastitis, both diseases that are particularly common in early lactation. Probably the commonality is caused by undernutrition at this time and does not indicate a direct relationship between the diseases. Hypocalcaemia has also been linked to retained placenta, which probably is a causal relationship, since low extracellular calcium concentrations reduce uterine muscle tone, as well as that of skeletal and rumen muscle tone, which produce milk fever and reduced appetite, respectively.

The susceptibility of cows to hypocalcaemia has a genetic component, with a greater incidence in Jersey and Guernsey cows and less in Ayrshire. Similarly, Swedish Red and White cattle have about double the incidence of the disease compared to Friesians. However, heritability values are variable and may be negatively correlated with milk yield, making it difficult to select cows for low susceptibility to hypocalcaemia.

There is an inverse relationship between a cow's age and blood calcium concentrations, with older cows having a much greater incidence of parturient paresis. Older cows have fewer receptors in both the gastrointestinal tract and bone for $1,25(OH)_2$ vitamin D and fewer osteoblasts and osteoclasts, leading to a reduced ability to mobilize calcium in early lactation. They also increase to peak lactation rapidly, thereby increasing the drain on calcium reserves.

The rapid demand for calcium at the start of lactation therefore places the high-yielding cow perilously close to metabolic breakdown. Greater attention must be given in future to this early lactation period, selecting cows with a gradual build-up to peak lactation, followed by a long lactation, including as long a period as possible of steady-state production, where nutrient demands are met by feed intake. Failure to address the early lactation problem will lead to increased incidence of diseases such as hypocalcaemia, with adverse effects on the welfare of the cow.

Magnesium
Magnesium is not regulated homeostatically to the same extent as calcium – a clear indication that over the period of evolution of mammalian mineral metabolism, magnesium deficiency was not a serious problem. It is, however, a mineral that can easily be deficient in lactating cows grazing intensively

managed pastures. Some magnesium can be mobilized from bone tissue during a deficiency, but this is unlike calcium in that it is not under the direct control of vitamin D. However, during hypomagnesaemic tetany a lactating cow will considerably reduce milk magnesium content, in just the same way that a cow with milk fever reduces milk calcium content. The acute form of the disorder produces initial signs of increased nervousness, with staring eyes and ears pricked. Muscles twitch and the animal may stagger in an uncoordinated fashion. When the animal becomes recumbent her legs peddle and she grinds her teeth, after which a coma usually sets in and she dies (Blowey, 1986). A chronic form of hypomagnesaemia also exists, with a gradual loss of condition and a stiff gait. The incidence of hypomagnesaemic tetany in dairy cows varies from 3 to 10% in developed countries.

One of the main risk factors for the disease is potassium fertilizer application during spring, since young grass has a low level of available magnesium. Both the magnesium concentration is low and the availability of the magnesium entering the rumen. The rumen is the main site of absorption, and if the herbage has a high potassium concentration the rapid efflux of K^+ ions into the blood stream creates a significant electrochemical potential difference across the rumen epithelial wall, with the rumen contents becoming negative relative to the blood stream. This hinders the absorption of divalent cations, especially magnesium. It can be countered by increasing the sodium content of the rumen contents, which inhibits the absorption of potassium by the animal (and has the same effect in controlling potassium absorption at the soil:root interface of plants). Increasing the plant's sodium concentration can be most easily achieved by reducing the application of potassium fertilizer and replacing it by sodium fertilizer. Some metabolic processes in the plant specifically require potassium, but about one-half of the total potassium used by the plant can be replaced by sodium without loss of plant yield. Sodium fertilizer has an additional benefit in dry weather, as it increases plant turgor by its osmotic effect in the extracellular compartments of the plant.

Nitrogen fertilizer is another risk factor for hypomagnesaemia, since a surplus of ammonium ions alkalinizes the rumen contents, which reduces the dissociation of magnesium ions and hence magnesium availability.

Above 18–20 mg Mg l^{-1} of blood serum, excess magnesium is excreted in the urine, and since magnesium absorption is not actively controlled, this can be taken as the threshold below which hypomagnesaemia is likely to occur. Magnesium availability from feed is particularly variable, ranging from 15–50%. In spring grass the lower values are more appropriate, but some magnesium supplements (oxides, carbonates and sulphides) are absorbed to a much greater degree (50–60%). Such supplements can be included in the diet of cows grazing spring grass or they can be spread directly on the pasture. This is time consuming and regular tractor passage is undesirable as it can damage the soil structure. Also, if alternative pasture is available, cattle will normally avoid the treated pasture. A soluble form of magnesium can be added to the drinking water in troughs, but care should be taken if cows are also using streams or other alternative drinking water sources. Also, during wet weather, which is a common risk factor for hypomagnesaemia caused by the reduced DM intake of wet grass, the animal's water intake from troughs will be reduced because of intake of water on the surface of the herbage. The classical conditions for hypomagnesaemia are therefore cold, wet conditions soon after the cows have been turned out to pasture. However, the magnesium content of pasture is variable and on many farms, the cows never experience hypomagnesaemia problems. Ensuring a high soil pH is important since this increases herbage magnesium content. The most sensitive indicator of hypomagnesaemia risk is the K:(Ca + Mg) ratio in herbage, which should be less than 2.2. It is likely that in future more attention will be paid to herbage mineral composition in grass breeding programmes.

Magnesium-enriched concentrate feeds can be fed to cattle after they have been turned out to pasture, but this is an expensive way of controlling the disease if it is the sole reason for feeding the concentrate supplement. The substitution rates of concentrate for grass soon after turnout is likely to be approximately 1:1, resulting in no nutritional benefit to feeding concentrate, other than increased magnesium intake. Many dairy farmers now rely entirely on complete diets for feeding their cows and do not feed the cows in the parlour, making it difficult to feed small quantities of concentrate supplements to grazing cows. Magnesium bullets are probably the safest option in a high-risk area, even though they can occasionally be expelled from the rumen during the reverse peristalsis in rumination. The bullets dissolve over a 2-month period, which should be sufficient to provide cover during the period after turnout.

The treatment of clinical cases is by administering magnesium solution either subcutaneously, or in emergencies by slow intravenous injection, followed by sedatives to keep the animal calm and supported to prevent muscle damage. Later the animal should be drenched with 70 g calcined magnesite. Too often, the disease is not diagnosed until it is too late.

Magnesium requirements can be estimated factorially as 18 mg Mg kg^{-1} live weight per day for maintenance (endogenous losses) + 2.7 g kg^{-1} live weight gain + 0.74 g Mg l^{-1} of milk + 3 g Mg day^{-1} for the last 8 weeks of pregnancy. For a typical cow giving 30 l day^{-1} of milk, the total magnesium requirement would be about 28 g day^{-1}. Since herbage often contains as little as 1 g Mg kg^{-1} DM and a cow may only eat 13 kg day^{-1} DM in inclement weather, the potential for deficits to occur is clear.

A MAGNESIUM BALANCE PROBLEM   Dairy heifers of 500 kg mean live weight and mean milk production of 30 l day^{-1} are grazing pasture with a mean Mg concentration of 2 g kg^{-1} DM. It is expected that they are consuming about 15 kg day^{-1} DM each, with no supplements, and that they will gain weight at 0.6 kg day^{-1}. Calculate their Mg balance and decide whether supplementation is required. If so, consider how this might best be achieved.

SOLUTION   Magnesium intake is 30 g day^{-1}. Requirements for endogenous losses are 8.8 g day^{-1}, for growth 1.6 g day^{-1} and for milk production

22.2 g day^{-1}, i.e. a total of 32.6 g day^{-1}. Intake can be supplemented by magnesium in feed or water, or calcined magnesite spread on the pasture. Magnesium bioavailability can be increased by reducing the potassium or increasing the sodium contents of herbage.

Sodium and potassium

The close relationship between these two elements makes it impossible to consider them in isolation. Many cattle, particularly high-yielding dairy cattle, do not receive enough sodium from their forage diet to compensate for the losses from the body. These are principally in urine, sweat and, in lactating cows, milk. By contrast potassium deficiencies do not occur in cattle that are fed a reasonable amount of forage. Forages have a major requirement for potassium in order to maximize growth, but only rarely do they have a specific requirement for sodium. However, cattle have a major need for sodium and will selectively graze areas of pasture that have been dressed with sodium compared to those that have not. It is quite simple to include sodium compounds, e.g. common salt, in a blended mixture of fertilizers, and cattle grazing sodium-enriched herbage will have an increased intake. The low cost of sodium chloride and, to a lesser extent, sodium nitrate, make them suitable for this purpose. Clinical sodium deficiencies occur at about 1 g Na kg^{-1} DM or below and are typified by depressed appetite, pica, low growth rates and milk yields and in extreme cases by collapse and possibly death. Many tropical forages contain less than 1 g Na kg^{-1} DM, and requirements will be increased if the cattle are in an environment where they sweat more (Sanchez *et al.*, 1994).

The sodium content of pasture is variable and dependent on the soil type, the species making up the sward and the amount of potassium added. It will usually be highest in pastures near the sea. The average content is about 2 g kg^{-1} DM in the UK, but the optimum for dairy cows is nearer 5 g kg^{-1} DM (Phillips *et al.*, 2000). This level of sodium in the sward can be achieved by the application of about 50 kg ha^{-1} year^{-1} sodium as a fertilizer, depending on conditions. The grass sodium:potassium ratio is the best guide to the quantity of sodium to apply. Applying too much sodium is unlikely to increase the sodium content of the grass much above 5–7 g kg^{-1} DM. However, if grass sodium does increase above 7 g kg^{-1} DM, as might occur if the grassland is close to the sea, then the digestibility of this grass is likely to be reduced.

In addition to preferring sodium-fertilized pastures, cows also spend more time grazing (Table 3.4) and eat the pasture with a faster biting rate. They then spend longer ruminating on the herbage and this, coupled with a decrease in the rumen acidity caused by recycled sodium in the saliva as a pH buffer, increases herbage digestibility and promotes the growth of acetogenic (fibre-digesting) bacteria in the rumen.

The high intake of grass that has been fertilized with sodium increases the cow's milk yield and weight gain (Table 3.4). The decrease in rumen acidity is particularly beneficial for cows grazing lush pasture with no forage supplements, when it may be reduced to below pH 6. Under such conditions milk fat content has been observed to increase from 37 to 40 g l^{-1}.

Table 3.4. The effects of using sodium fertilizer on grass for grazing dairy cows in temperate conditions.

	No sodium	Sodium
Grass sodium (g kg^{-1} DM)	2.0	4.0
Grass magnesium (g kg^{-1} DM)	1.8	2.0
Grass calcium (g kg^{-1} DM)	4.5	4.9
Milk yield (l day^{-1})	18	20
Butterfat (g kg^{-1})	37	40
Cow weight gain (kg day^{-1})	0	+0.3
Grazing time (h day^{-1})	8.7	9.6

Potassium fertilizer can restrict the uptake of magnesium and calcium by herbage plants. Sodium fertilizer can reduce potassium uptake and thereby increase herbage magnesium and calcium absorption. Potassium is the more efficient of the two monovalent cations in activating some plant enzymes, but its adverse effects on the absorption of other minerals in the ruminant make it advisable for its use to be restricted, particularly at critical times of year when lush pasture is available. Less potassium and more sodium in the rumen also increases the ruminal absorption of magnesium and calcium, which means that grass staggers and milk fever are less likely to occur. Although increased magnesium absorption in cattle with sodium fertilizer application to herbage has been demonstrated recently, it was first suggested over 30 years ago, when Paterson and Crichton (1964) found that low sodium intakes were responsible for the high level of grass staggers in some British dairy farms. Magnesium is also involved in the cow's immune system and some reductions in milk somatic cell counts in cows grazing sodium-fertilized pastures have been observed.

Sodium also helps the plant to make best use of available sulphur, which is another element that potentially limits grass growth. Farmers should not rely on applying only nitrogen, phosphorus and potassium to their cattle pastures, but should apply a balanced cocktail of minerals to feed the grass plant. It is essential to consider the effects of fertilizers on the composition of herbage when planning the fertilizer regime. The grass plant must be balanced for minerals, such as sodium, magnesium and calcium. It is often easier and always safer to ensure that the grass eaten by the cattle has the right composition, rather than trying to correct deficiencies by adding minerals to the concentrate part of the diet. However, in some cases it may be wasteful to apply minerals to the land as some of them may be lost by leaching. In rangeland conditions, fertilizer application is not feasible, and salt licks are widely used to ensure adequate sodium intake. It is assumed that the cattle will consume to their requirements and no more. It is not easy to determine whether this is achieved, but the wide variation in attendance at salt licks suggests that other factors, such as social pressure, partly determine the intake of sodium in these circumstances.

Sodium supplements are beneficial for young milk-fed calves, where the stress of confinement may impair kidney function and increase sodium requirements. Much of the licking and other oral stereotyped behaviours seen in individually housed calves may be caused by sodium or other mineral deficiencies. Sodium supplements provided to the young calf will condition their sodium appetite and they will have greater sodium appetites as older cattle (Phillips *et al.*, 1999).

The sodium intake of cattle is closely related to their water intake, as water is required to maintain the osmotic balance. When sodium supplements are included in a complete diet or the sodium content of forage is increased, as occurs when sodium fertilizers are used or forages are upgraded or preserved with sodium-containing chemicals, adequate water must be available to the animals. The excessive urination of cattle fed conserved feed treated with sodium may cause difficulties in maintaining a clean environment for the cattle.

A SODIUM MINERAL BALANCE PROBLEM   Early lactation dairy cows with a mean weight of 650 kg and giving 35 l day^{-1} of milk are grazing pasture with a mean herbage sodium concentration of 2 g kg^{-1} DM. They are expected to consume about 15 kg day^{-1} DM. Calculate their sodium balance and discuss whether supplementation is required and how this would be best achieved.

SOLUTION   The cows' sodium intake is 30 g day^{-1}. Their requirement is 5.2 g day^{-1} for maintenance and 22.4 g day^{-1} for milk production.

POINTS TO CONSIDER

- Is herbage sodium concentration accurate?
- Does herbage mineral concentration affect intake?
- Is the ratio of sodium to other minerals important for animal health?
- What is the sodium intake from drinking water or other feeds?
- Are the temperatures sufficiently high to make the cattle lose extra sodium through sweating?

Phosphorus

Phosphorus is required for the bone matrix, for energy transfer in ATP and ADP and for phospholipid membranes. The absorption is mainly in the small intestine and is actively controlled by vitamin D. In areas where cattle are housed for long periods of the year, the low vitamin D status of the cattle will increase phosphorus requirements, unless supplementary vitamin D is fed. Absorption declines markedly with age, from nearly 100% in the suckled calf to 43% for cattle over 1 year of age. There is some evidence that a very high calcium:phosphorus ratio (>10:1) in the feed reduces phosphorus availability. The optimum ratio of calcium to phosphorus is probably between 1:1 and 2:1, since this is the ratio in bones. In many feeds most of the phosphorus is in the form of phytate, which is readily available to ruminants but not monogastric animals, because the rumen microorganisms have the ability to hydrolyse the phytate phosphorus with the enzyme phytase. Most low-quality feeds fed to beef or dry dairy cows contain little phosphorus, since this declines as the plant matures. Problems are most likely, therefore, to occur in cattle on rangelands, where the soil phosphorus status is low and the forage is mature (Ternouth, 1990).

Most excess phosphorus is excreted in faeces, and environmental concerns now dictate that more effective phosphorus rationing is utilized for dairy cattle in particular (Berentsen and Giesen, 1997). Phosphorus losses can be reduced by feeding a diet with a reduced phosphorus concentration when milk yields decline in mid–late lactation. Some farmers feed a diet with a higher concentration of phosphorus in early lactation as there is evidence that phosphorus is important during the service period. However, there is little scientific evidence of any benefit of supplementary phosphorus in early lactation, except in the situation where low phosphorus contents of the feed reduce intake and the cows become very thin.

Phosphorus deficiency is manifested as a stiff gait, bone abnormalities and non-specific conditions such as reduced production (growth rate or milk production), low feed intakes and a depraved appetite, or pica. The latter is not peculiar to phosphorus, but causes animals to search for abnormal feed sources, such as bones or soil, which may restore the mineral deficit. Bone chewing in areas where putrified carcasses are contaminated with bacteria such as *Clostridium botulinum* can be dangerous. Blood plasma inorganic phosphorus contents can be used to predict whether cattle have adequate phosphorus supply. They should be in excess of 4–6 mg dl^{-1}. Requirements are difficult to estimate because absorption varies widely. If it is assumed to be 60%, most rationing systems suggest that a dairy cow will require about 24 g day^{-1} for maintenance and 1.5 g l^{-1} for milk production, or a phosphorus content in the ration of about 0.5% for a high-yielding cow. Growing cattle are likely to require a phosphorus concentration in the ration of about one-half of this value. If phosphorus intakes are inadequate, supplements can be mixed with salt and molasses and provided as field blocks, although cattle are not able to regulate their intake very accurately in relation to requirements. This is probably the best supplementation method for range cattle. In more intensive grazing conditions, the phosphorus contents of herbage can be increased by phosphorus fertilizer quite effectively. Some phosphorus supplements, such as dicalcium phosphate or superphosphate, are well absorbed, but rock phosphate is not well absorbed, is unpalatable and may contain a high fluoride content.

Sulphur

Sulphur is required mainly for sulphur-containing amino acids and as a result of this, requirements are often stated in relation to protein, or rumen-degradable protein, requirements. The ratio of S:N in microbes, milk and tissue is approximately 0.07:1, which should therefore be the basis for sulphur requirements. Sulphur is a particularly important macroelement for cattle because the range between deficiency and toxicity is quite narrow.

For most of the 20th century, the sulphur intake of grazing cattle in the industrialized countries has been increased by the sulphur deposits from emissions to the atmosphere by power stations and other heavy industry, and the use of low-grade fertilizers, which have high sulphur contents. More recently, emissions have decreased and sulphur availability in the soil has been demonstrated to limit grass growth, especially since the application of basic slag ended. Sulphur fertilizers acidify the upper horizons of the soil and have been used to counteract salinity. Grass crops that are heavily fertilized with nitrogen also need sulphur fertilizer, otherwise they will have a reduced content of true protein and more non-protein nitrogen. However, although the grass crop increases its growth rate in response to sulphur fertilizer, the sulphur concentration may exceed the toxic threshold for cattle (approximately 2 g S kg^{-1} DM) with heavy sulphur applications. Acetic acid production in the rumen is then reduced, which can reduce milk fat production. Also, excessive sulphur application to pasture reduces the herbage content of several trace elements, such as copper, molybdenum and boron (Chiy *et al.*, 1999). However, in high molybdenum areas the reduction in copper content with sulphur may be outweighed by the increase in availability caused by reducing the molybdenum content of pasture. Sulphur toxicity can also occur if high sulphate molasses or distillers' solubles are fed, or if sulphur is added to the feed to control urinary calculi.

For cattle fed large quantities of non-protein nitrogen, sulphur deficiencies are likely unless supplements are fed. The critical level in feeds is approximately 1 g S kg^{-1} DM. Supplements of elemental sulphur are not as well utilized by the rumen microorganisms as compounds such as sodium sulphate.

Trace elements

Many elements are required only in very small quantities or traces. These are known as trace elements, but just because they are required in small quantities, it does not mean that a deficiency is unimportant. Toxicities are rare in cattle and are considered in Chapter 9.

As research progresses, the essentiality of more and more elements can be proven for cattle, although for many elements this is of little practical use since they are present in cattle feeds in concentrations well in excess of requirements. Many of the trace elements are bound to phytate in feed, but ruminants are much better than monogastric animals at obtaining the necessary cation as rumen microorganisms produce phytase to digest the phytate.

Copper
Copper is a component of many metallo-enzymes (especially cytochrome oxidase for energy metabolism, caeruloplasmin for iron transport in blood, superoxidase dismutase to destroy superoxide radicals that cause cellular

damage and tyrosine for melanin production). It is also an important anti-oxidant and is particularly used for blood formation. It is stored mainly in the liver, which can extract excess copper from caeruloplasmin.

The presence of several minerals will inhibit the absorption of copper, because of competitive inhibition. These include molybdenum, sulphur, iron, zinc, cadmium and possibly calcium. The complexity of the interrelationships between these elements makes the prediction of copper availability difficult. The true availability (taking endogenous losses into account) is considerably more difficult to determine than the apparent availability, since the endogenous losses, which are mainly in bile and pancreatic juices, are hard to quantify.

An acute copper deficiency occurs commonly in areas with high molybdenum contents of soil and pasture. Copper, molybdenum and sulphur combine in the rumen to form an unabsorbable complex, copper thiomolybdenate, which is excreted in faeces. Iron supplementation or iron consumed in soil, or silage contaminated with soil, will exacerbate the problem.

The main symptoms of copper deficiency are scouring, anaemia (because of reduced iron absorption) and weak bone formation. Ataxia occurs in calves (commonly known as sway-back), and in adult cattle (known mainly as falling disease). Despite the recent increase in use of sulphur fertilizer on British grassland, which many feared would increase copper deficiency in cattle, the incidence of the disease has not increased, probably because sulphur fertilizers reduce herbage molybdenum content. High sulphur contents in drinking water, however, will reduce copper availability (Smart *et al.*, 1992). Copper deficiency, or hypocupraemia, can be recognized by its effects on hair coloration, particularly around the eyes where the loss of pigmentation causes dark-coated animals to have a bespectacled appearance. The coat also loses its sheen and the copper contents of the hair can be used to indicate the copper status of an animal. Another diagnostic method is to determine the blood copper content, which should not be less than 80–100 μg per 100 ml for whole blood or 60 μg per 100 ml for blood plasma. Copper deficiencies can cause a reduced growth rate in cattle in susceptible areas.

Herbage copper contents of 5–8 mg kg^{-1} DM should be sufficient if there are no complications of, for example, a high molybdenum content, as in the 'teart' pastures of south-west England. However, the maturity of the herbage will affect its availability, being sometimes 1% in lush autumn herbage but 7% in hay. In autumn, the quantity of herbage available to cattle is often low and soil will be consumed as cattle graze close to the soil surface, with adverse effects on copper absorption of the iron consumed in the soil. Copper compounds should be added to mineral supplements in areas where cattle have been reported to suffer from hypocupraemia, to increase the dietary copper content to approximately 10 mg kg^{-1} DM. Alternatively copper can be provided in slow-release boluses or needles, which are administered to the rumen, from where some copper will be absorbed, with the majority being absorbed in the small intestine.

Copper toxicity is rare in adult cattle as a result of the low absorption rate. They will tolerate herbage containing up to 100 mg Cu kg^{-1} DM, but can be poisoned by eating soil that has been contaminated with copper, or by consuming diets where there has been an error in mineral formulation. If the diet has a low molybdenum concentration, feeds with normal copper content can induce copper toxicity. Sheep are, however, much more susceptible. Copper toxicity can occur in calves, as absorption is much greater at this age – up to 60–80% in the first few weeks of life. As the animal grows the absorption rate decreases rapidly, particularly when it develops a functional rumen. The susceptibility of sheep to copper toxicity has resulted in maximum copper inclusion rates in ruminant feeds used in the EU of approximately 35 mg kg^{-1}.

The symptoms of copper toxicity are nausea, vomiting, abdominal pain, convulsions, paralysis, collapse and death. It can be treated by administering molybdenum and sulphate, usually as a drench of ammonium molybdenate and sodium sulphate.

Iron

Iron is required for the formation of haemoglobin, the body's major oxygen carrier, and for oxidizing catalysts. It is also present in blood as ferritin and transferrin, and in muscles as myoglobin. Milk has a particularly low concentration of iron, in relation to other minerals, but nature attempts to overcome iron deficiencies in the suckling period by providing a store of iron at birth in the liver (about 450 mg). The restriction of iron availability to bacteria limits their growth in the mammary gland and thus helps to prevent mastitis. Lactoferrin binds iron and thus controls its release in milk. It is particularly present in the milk residues in non-lactating mammary glands and in mastitic milk, demonstrating that it has a protective role of limiting iron availability for bacterial growth.

Iron deficiency is only found in milk-fed calves and may be deliberately induced to produce pale meat, which consumers associate with tenderness. The deficiency is typified by anaemia, reduced appetite, poor growth rates and inability to cope with exercise. Iron also plays a part in the formation of immunoglobulins, so that iron-deficiency anaemia is associated with reduced immunocompetence. Iron is very effectively recycled within the body and there is little urinary excretion. Iron absorption is enhanced when cattle are iron deficient, but the milk-fed calf will still develop anaemia within 8–12 weeks unless iron supplements are provided. Veal calf producers aim to create low muscle myoglobin contents without adversely affecting appetite or growth rate. This requires approximately 25–50 mg Fe kg^{-1} of dietary DM, although iron availability is subject to the same uncertainties as copper. However, milk contains only 5 mg l^{-1} Fe, and the stores provided to the calf at birth are therefore vital. Calves that suckle their dams at pasture invariably start consuming a few leaves of grass before the anaemia becomes established and therefore are not at risk. The iron content of leaves is much greater than seeds or milk and iron deficiency does not occur in cattle fed forage-based diets. Indeed an iron supplement should not be added to the diet of adult cattle at risk of hypocupraemia, because it will inhibit the absorption of copper.

The iron status of calves can be assessed from the haemoglobin content of blood. The critical level is 4.5 mmol l^{-1}, at which level the muscle myoglobin content will be reduced, but not growth rate or the animal's ability to exercise. Legally, in the EU, all calves must be fed sufficient iron to ensure a blood haemoglobin level of at least 4.5 mmol l^{-1}. If blood haemoglobin levels are less than this, then a supplement of between 25 and 50 mg Fe kg^{-1} should be provided. Providing solid feed as required by EU legislation will also help to prevent low blood haemoglobin levels. In the long term, consumers should be made aware that veal meat that is red is just as good quality as white meat and will indicate that the calves have not suffered impaired welfare as a result of hypoferraemia.

Cobalt

Cobalt is utilized in the rumen and it is an essential constituent of vitamin B_{12} and its analogues. Deficiencies are common where soils have low cobalt concentrations and are known by a variety of local terms, such as 'pine'. Excessive liming is a risk factor in reducing cobalt availability to plants. A depressed appetite and low growth rates are the symptoms in the early stages of the disease, followed by muscular wasting, pica and severe anaemia. Pasture should contain at least 0.1 mg Co kg^{-1} of DM and cattle deficiencies can be most accurately diagnosed from liver cobalt or vitamin B_{12} contents. The liver stores surplus cobalt in the form of vitamin B_{12}, and the mean cobalt concentration should not be less than 0.06 mg kg^{-1} DM. A specific instance of cobalt deficiency has been observed in cattle in Australia and New Zealand grazing pasture containing *Phalaris tuberosa*. The plant contains a neurotoxin, N,N-dimethyltryptamine, which is inactivated by rumen microorganisms in the presence of adequate cobalt. If there is inadequate cobalt in the rumen, cattle develop an ataxia condition known as 'Phalaris staggers' (Clement and Forbes, 1998).

Deficiencies can be rectified by using cobalt-containing fertilizers in susceptible areas. If soils are alkaline the uptake by the plant is low and direct supplementation of the cattle is necessary, by including cobalt in salt licks, slow-release boluses or even by regular drenching.

Selenium

Selenium is an important component in the cell enzyme, glutathione peroxidase, which controls the peroxides in the cytosol that react with unsaturated lipids to cause cell damage. The same control can also be achieved in membranes by tocopherol, vitamin E, which directly inhibits the auto-oxidation of polyunsaturated fatty acids by oxygen metabolites, hence the same symptoms are exhibited for selenium and vitamin E deficiencies.

Selenium deficiency causes nutritional muscular dystrophy, which is known as white muscle disease in calves. In extreme cases the heart muscle degenerates and myoglobin may be released to give the urine a red coloration.

The deficiency is widespread where soil selenium concentrations are low, producing feeds with selenium concentrations less than 0.05 mg kg^{-1} DM. Such low concentrations are common, with approximately two-thirds of dairy cows in the USA and Europe being in areas where the soils are selenium deficient. Supplementation is normally provided in mineral mixes in the form of sodium selenite to increase the selenium content of the diet up to 0.1 mg kg^{-1} DM. For cattle that do not regularly receive mineral supplements added to concentrate feeds, such as those on rangeland, selenium can be added to salt licks or given by injection. The benefits of selenium and vitamin E supplementation are not additive, as both achieve the same detoxifying effect, albeit in different parts of cells. However, supplementation with one will have a sparing effect on the requirement for the other.

Care must be taken when planning selenium supplementation for cattle, as toxicity can also occur. There are many areas where selenium concentrations in plants, particularly accumulator plants, are in excess of 5 mg Se kg^{-1} of DM, which is the threshold for toxic symptoms in cattle. Toxicity will depend largely on the extent to which cattle consume the accumulating plants, which may in turn depend on the availability of other herbage.

Selenium is also important in the immune system, as the respiratory burst by phagocytes during an infection increases oxygen metabolism, resulting in proliferation of hydrogen peroxide and superoxide. The free radicals cause cell damage and limit the bactericidal effectiveness of the neutrophil burst. Supplementation with selenium can therefore help to control mastitis in dairy cows. Current evidence suggests that it is beneficial to increase selenium intake for early lactation cows up to 4 mg day^{-1}, which is about twice the dietary concentration recommended to avoid muscular dystrophy (0.1 mg kg^{-1} DM). Herd selenium status can be monitored from blood samples, and is adequate when the concentration in whole blood is between 0.2 and 1 μg ml^{-1}.

Zinc

Zinc is one of the commonest metals in enzyme complexes in biological systems, most notably in DNA and RNA synthesis and protein metabolism. Deficiencies are confined mainly to areas with zinc-deficient soils. A variety of symptoms are observed, but impairment of growth and reproduction are most common. Part of the reduction in growth is caused by lower appetite, but part is because of impaired protein metabolism. Disorders of the integument are also regularly observed – parakeratosis in calves and hoof disorders, such as pododermatitis, in adult cattle. The involvement of zinc in protein metabolism is important for keratin deposition, firstly in hooves but also in the teat canal, where it is a primary barrier to mammary infections. Zinc deficiencies reduce the incorporation of several amino acids into skin proteins. Zinc concentrations are normally high in body hair and the gonads, and it is here that zinc concentrations decline most rapidly during a deficiency. The essentiality of zinc for DNA and RNA synthesis leads to impairment of T lymphocyte proliferation during infection in zinc-deficient cattle, but other components of the immune response are not seriously affected. Inverse relationships between zinc status and somatic cells in milk have been observed but the exact cause is not yet known.

Zinc is stored in a number of tissues, notably the liver and bones. In the liver, it is complexed by metallothioneins that function both to store surpluses and absorb toxic quantities that may be consumed. A deficiency is best determined from blood plasma concentrations, since a large and rather immutable quantity of the body's zinc is contained in the erythrocytes. The critical level of plasma zinc is 0.4–0.6 mg l^{-1}. Requirements are difficult to state precisely, mainly because of the extensive interactions with other elements, in particular copper, calcium and cadmium. However, 30–40 mg Zn kg^{-1} of feed DM is generally recommended for most classes of cattle, unless the feed contains a high copper level, in which requirements will be greater. Because zinc toxicity is very rare in cattle (up to 1 g kg^{-1} DM can be tolerated), generous supplementation is usual in deficient areas. This can be by adding zinc compounds to salt licks, by zinc fertilizer, which may be required for plant growth anyway, or by direct addition to a complete diet.

Vitamins

Vitamin A

Vitamin A is required for the proper formation of epithelial and bone tissue, and in particular for the formation of retinol, an important component of scotopic vision. The daily requirements for vitamin A by growing cattle are about 66 IU kg^{-1} live weight, and for lactating cows about 40,000 IU for maintenance and 4000 IU l^{-1} for production. The precursors of vitamin A, carotenoids, exist in plant material but they are unstable, and preservation and drying of forages can greatly reduce the carotenoid concentrations. Although the conversion rate of the most common form – β-carotene – depends on many factors, to determine the supply from feed it can be assumed that calves and growing cattle produce 1 μg (3.3 IU) of vitamin A for each 6 μg of β-carotene. For lactating cows, the conversion is less efficient, being only 1 μg (3.3 IU) vitamin A per 32 μg β-carotene. Thus, a lactating cow consuming 15 kg herbage DM with a concentration of 100 mg β-carotene kg^{-1} of DM will produce 150,000 IU of vitamin A, well in excess of requirements. However, carotene concentrations are usually much less for conserved feeds – typically straw contains 5 mg β-carotene kg^{-1} of feed DM, hay 10–20 mg β-carotene kg^{-1} of feed DM, grass silage 120 mg β-carotene kg^{-1} of feed DM and maize silage 11 mg β-carotene kg^{-1} of feed DM.

A deficiency of vitamin A reduces the production of some reproductive hormones and rebreeding dairy cows may be difficult (Kolb and Seehawer, 1997). Milk yield may also be reduced. Vitamin A can be stored in the liver and released over a period of several months, if needed. Supplementation of dairy cows with 200–300 mg day^{-1} β-carotene is recommended if plasma β-carotene concentrations are below approximately 2500 μg l^{-1}.

Vitamin D

Vitamin D can be obtained by the irradiating action of sunlight on the skin, providing that there is not too much hair covering it, but cattle that are housed have to obtain vitamin D from their feed or from surpluses which have been stored in adipose tissue, to be released as required. As the concentrations are negligible in most feeds, supplementation is required, at a daily rate of 6 IU kg^{-1} live weight for growing cattle and 10 IU kg^{-1} live weight for adult cows. Compounded concentrate feeds usually contain 1000–2000 IU kg^{-1} of vitamin D, which is sufficient for most purposes. However, housed suckler cows that are only fed a small amount of concentrates daily will need a higher level of supplementation. Inadequate vitamin D intake causes osteoporosis and reduced fertility.

The conversion of vitamin D into its active form is regulated by plasma calcium and phosphorus concentrations and parathyroid hormone. Large doses of vitamin D fed to dairy cows around calving mobilize calcium from bone tissue and increase calcium uptake from the gut. They are quite effective in reducing the incidence of milk fever, with the recommended intake for this purpose being 40,000–70,000 IU day^{-1}.

Vitamin E

Vitamin E is an antioxidant, which preserves the integrity of cell membranes. In severe deficiencies, the myoglobin content of muscles is depleted and they turn white (White Muscle disease) (Rammell, 1983). Calves are more likely to suffer from vitamin E deficiency than older cattle, because they have few reserves. The standard vitamin E addition to milk replacer is α-tocopheryl acetate, which is particularly needed if unsaturated fatty acids are added to the milk. Soya, maize or palm oils are now commonly added to milk replacers to increase their energy value to the calf.

Cattle with deficient or marginal vitamin E and selenium status are more likely to have retained placentae and may have fertility disturbances. Vitamin E is important for udder health, with the provision of supplementary vitamin E and selenium having additive benefits in controlling mastitis. Whereas selenium and vitamin E tend to have synergistic effects, vitamins E and A are antagonistic. The requirements for vitamin E are 1–2 mg kg^{-1} live weight, provided that there is sufficient selenium in the diet.

The B vitamins

Thiamine (vitamin B$_1$) is synthesized by the rumen microorganisms but some of the feed thiamine is also denatured in the rumen. Most thiamine produced in the rumen is absorbed in the small intestine and should be sufficient for all but the highest yielding cows. However, there appear to be certain conditions when thiaminases are manufactured, such as during acidosis, and the thiamine produced may be inadequate. The conditions under which thiamine production is inadequate are not yet sufficiently understood to recommend supplementation of cattle. However, if a problem is suspected, a supplement of 10 g t^{-1} of compound feed can be added.

Another B vitamin, niacin, is also synthesized in the rumen, but may be deficient. Supplementary niacin has been demonstrated to increase nutrient digestion rate in the rumen of the lactating dairy cow, which can prevent high-yielding cows from developing ketosis. Approximately $3\,g\ day^{-1}$ is recommended for high-yielding dairy cows, from 2 weeks before calving to week 10 of lactation.

Vitamin B_{12} is endogenously synthesized in adequate quantities, provided that the diet contains sufficient cobalt. Vitamin B_{12} analogues are also synthesized, and if the diet contains too much concentrate feed, too many analogues are produced and too little active B_{12}. To rectify a deficiency, dietary supplements of either vitamin B_{12} or cobalt can be provided, in particular to high-yielding dairy cows.

Biotin is also synthesized in the rumen, usually in sufficient quantities for high milk yields. Some research has indicated that biotin supplements can reduce lameness by improving keratin production.

Water

The requirements of cattle for water are met by imbibed water, water in food and by the water produced by metabolic reactions in the animal's body. Requirements are influenced particularly by temperature, humidity, the nitrogen, sodium and DM contents of the feed and milk yield. High nitrogen and sodium intakes have to be excreted in the urine with the addition of water, hence the voluntary water intake increases to maintain osmolarity. Feeds of high DM content increase voluntary water consumption, as they require the addition of more saliva before they can be swallowed. Cows consuming mainly dried rations, such as hay and concentrates, therefore require more water. The water allocation to cattle can be divided into a requirement for maintenance, at $0.09\,l\ kg^{-1}$ body weight, and a requirement for milk production, $2–2.5\,l\ l^{-1}$ milk produced. A small reduction in water supply can be tolerated as there is some luxury uptake, but a severe restriction will lead to reduced milk yield. This may happen if water pipes freeze, and water intake will also be reduced if the water provided is too cold.

Lactating cows naturally drink four or five times per day and, if water is provided only at milking times, intake is likely to be restricted. Cows particularly like to consume water after being milked and after they have eaten, to restore their osmotic balance. Peak intake is likely to be in the evening, when there is a concentrated feeding period. The water supply should be clean and unpolluted. Allowing cows access to dirty streams to obtain their water is likely to spread disease. Usually water is provided in a trough, which should have a bar around it to prevent cows defecating in it. The troughs should be cleaned out regularly otherwise a sludge develops at the bottom which will reduce intake. In the field large concrete troughs are sometimes used as they store a lot of water. However, because the cows cannot reach to the bottom they need regularly cleaning out. The field troughs should be centrally situated so that cows do not have to walk far, and it is best to prevent wildlife from using them, for example badgers, because of the risk of disease transmission.

Indoor troughs are usually smaller than those in the field and need careful siting. The end of a cubicle row is suitable, but care should be taken that the floor does not become slippery around the trough, as cows will be making tight turns there and may slip over. If the water trough is sited in the feeding passage, feed may enter it when it is delivered from a mixer wagon or forage box. Shallow troughs that fill rapidly are best, as the water that they provide for the cow is usually clean and fresh.

References

Agricultural and Food Research Council (AFRC) (1992) Technical committee on response to nutrients. Report No. 9. Nutritive requirements of ruminant animals: protein. *Nutrition Abstracts Review* (Series B), 62–71.

Agricultural Research Council (ARC) (1980) *The Nutrient Requirements of Ruminant Livestock.* Commonwealth Agricultural Bureaux, Slough.

Alderman, G. and Cottrill, B.R. (1993) *Energy and Protein Requirements of Ruminants – An Advisory Manual Prepared by the AFRC Technical Committee on Responses to Nutrients.* CAB International, Wallingford.

AUS (1990) *Feeding Standards for Australian Livestock: Ruminants.* Standing Committee on Agriculture, Ruminant Sub-Committtee (Corbett, J.L., ed.). CSIRO, Melbourne.

Ausschuss für Bedarfsnormen (1986) *Energie und Nährstoffbedorf Landwirtschaftlicher Nutztiere Nr 3. Milchkühe und Aufzuchtrinder.* Forschungsinstitut für die Biologie Landwirtschaftlicher Nutztiere, Dummerstorf, Germany.

Berentsen, P.B.M. and Giesen, G.W.J. (1997) Potential consequences at farm level of the use of different policy instruments to decrease N and P_2O_5 surpluses on Dutch dairy farms. In: Sorensen, J.T. (ed.) *Livestock Farming Systems, More than Food Production.* EAAP Publication No. 89. Wageningen Pers., Wageningen, pp. 324–328.

Blowey, R.W. (1986) *A Veterinary Book for Dairy Farmers.* Farming Press, Ipswich.

Butler, W.R. (1998) Review: effect of protein nutrition on ovarian and uterine physiology in dairy cattle. *Journal of Dairy Science* 81, 2533–2539.

Central Veevoederbureau (1991) *Veevoedertabel.* Central Veevoederbureau in Nederland, Lelystad.

Chiy, P.C., Avezinius, J.A. and Phillips, C.J.C. (1999) Sodium fertilizer application to pasture 9. The effects of combined or separate applications of sodium and sulphur fertilizers on herbage composition and dairy cow production. *Grass and Forage Science* 54, 312–321.

Clement, B.A. and Forbes, T.D.A. (1998) Toxic amines and alkaloids from Texas acacias. In: Garland, T. and Barr, A.C. (eds) *Toxic Plants and Other Natural Toxicants.* CAB International, Wallingford, pp. 351–355.

Institut National de Recherche Agronomique (INRA) (1989) *Ruminant Nutrition. Recommended Allowances and Feed Tables* (Jarrige, R., ed.). INRA, Paris.

Kolb, E. and Seehawer, J. (1997) The significance of carotenes and of vitamin A for the reproduction of cattle, of horses and of pigs – a review. *Praktische Tierarzt* 78, 783.

Landis, J. (1992) Die Protein- und Energieversorgung der Milchküh Schweizerische. *Landwirtschaftliche Monatsschrift* 57, 381–390.

Madsen, J. (1985) The basis for the Nordic protein evaluation system for ruminants. The AAT/PBV system. *Acta Agriculturae Scandinavica* 25 (Supplement), 103–124.

Ministry of Agriculture, Fisheries and Food, Department of Agriculture and Fisheries for Shetland and Department of Agriculture for Northern Ireland (1975) *Energy Allowances and Feeding Systems for Ruminants.* Technical Bulletin 33. HMSO, London.

Paterson, R. and Crichton, C. (1964) Low level feeding of concentrates to dairy cows. *Veterinary Record* 76, 1261–1274.

Phillips, C.J.C., Youssef, M.Y.I., Chiy, P.C. and Arney, D.R. (1999) Sodium chloride supplements increase the salt appetite and reduce sterotypies in confined cattle. *Animal Science* 68, 741–748.

Phillips, C.J.C., Chiy, P.C., Arney, D.R. and Kärt, O. (2000) Effects of sodium fertilizers and supplements on milk production and mammary gland health. *Journal of Dairy Research* 67, 1–12.

Rammell, C.G. (1983) Vitamin-E status of cattle and sheep. 1. A background review. *New Zealand Veterinary Journal* 31, 179–181.

Sanchez, W.K., McGuire, M.A. and Beede, D.K. (1994) Macromineral nutrition by heat-stress interactions in dairy-cattle – review and original research. *Journal of Dairy Science* 77, 2051–2079.

Smart, M.E., Cymbaluk, N.F. and Christensen, D.A. (1992) A review of copper status of cattle in Canada and recommendations for supplementation. *Canadian Veterinary Journal - Revue Veterinaire Canadienne* 33, 163–170.

Sniffen, C.J., O'Connor, J.D., Van Soest, P.J., Fox, D.G. and Russell, J.B. (1992) A net carbohydrate and protein system for evaluating cattle diets: II. Carbohydrate and protein availability. *Journal of Animal Science* 70, 3562–3577.

Ternouth, J.H. (1990) Phosphorus and beef-production in Northern Australia. 3. Phosphorus in cattle – a review. *Tropical Grasslands* 24, 159–169.

Webb, K.E., Dirienzo, D.B. and Matthews, J.C. (1993) Symposium: nitrogen-metabolism and amino-acid nutrition in dairy-cattle – recent developments in gastrointestinal absorption and tissue utilization of peptides – a review. *Journal of Dairy Science* 76, 351–361.

Further Reading

Agricultural and Food Research Council (AFRC) (1992) Technical committee on response to nutrients. Report No. 9. Nutritive requirements of ruminant animals: protein. *Nutrition Abstracts Review* (Series B), 62–71.

Agricultural Research Council (ARC) (1980) *The Nutrient Requirements of Ruminant Livestock.* Commonwealth Agricultural Bureaux, Slough.

Alderman, G. and Cottrill, B.R. (1993) *Energy and Protein Requirements of Ruminants – An Advisory Manual Prepared by the AFRC Technical Committee on Responses to Nutrients.* CAB International, Wallingford.

Underwood, E.J. and Suttle, N.F. (1999) *The Mineral Nutrition of Livestock,* 2nd edition. CAB International, Wallingford.

Grazing Management and Systems 4

Introduction

Cattle evolved as grazing animals. Their form, behaviour and temperament are adapted to the grazing lifestyle. However, the consumption of grass alone cannot supply the nutritional requirements of very high yielding cows, neither does it always make the most efficient use of limited land stocks. Thus we have seen the recent evolution of permanent housing systems, for example, in parts of the world where high yields are required, such as the USA, also in areas where land is not suited to the production of grass, such as much of Israel, and finally where land is in short supply, such as in the Nile delta of Egypt. Grazing systems have continued to be profitable in areas where milk yields are modest, no more than approximately 5000–7000 l year^{-1} per cow, and grass grows throughout most of the year, such as in New Zealand, Eire and the western parts of the UK. The environmental conditions required for successful grazing systems are a high rainfall that is evenly distributed throughout the year and a mild climate. In the UK, there are 7 million ha of grassland in a total agricultural area of 20 million ha, and 6 million ha of this is rough grazing. It is therefore not surprising that most cattle are kept on the poorer quality land and that grass is the most important food for cattle, contributing an average of 75% of their metabolizable energy (ME) intake.

The feeding of conserved food to cattle has the advantages that the quality can be carefully controlled. Also housing the cattle to feed them means that they are close at hand for observation, treatment and milking. The main disadvantage is that conserved food is more expensive than grazed grass, and cows tend to stay healthier when they are outside. From the public's point of view, the grazing dairy cow is less stressed than her housed counterpart, she has room to move, companionship and fresh air. This is an anthropomorphic assessment, but we do not have sufficient evidence of the requirements of cattle to make other judgements.

Breeding for increased efficiency of feed utilization by cattle is likely to mitigate against grazing systems. Cattle are less efficient at harvesting a high proportion of the grass than a mowing machine. An average of 65% or less of the grass that is grown is harvested by cattle compared to 75% for silage making. Also, the animal's maintenance energy requirements are typically increased by 25% in grazing systems, compared with housed feeding. However, recent research suggests that the cow uses the period at pasture to recuperate from winter housing and permanent housing may therefore compromise the welfare of dairy cows in some intensive systems (Krohn and Rasmussen, 1992). Permanent housing will be encouraged by increasing possibilities for mechanization of the dairying systems, e.g. robotic milking, and by continuing escalation of labour costs.

The inadequacy of grazed herbage for cows yielding more than about 25 l day^{-1} of milk means that supplements have to be fed if cows are to remain in good condition. Typically, the high energy and protein content of grazed grass in spring encourages the cow to produce high milk yields with high protein and low fat contents. However, the low intake rate of grazed grass, about 20 g min^{-1} dry matter (DM) compared with twice that value for conserved forages, means that intakes are inadequate and energy requirements have to be partly met by catabolizing body fat tissue. Thus the cow loses condition and towards mid-summer, at a time when grass often is reduced in availability, the digestibility of the grass falls and milk yield starts to decline considerably unless supplements are fed. The spring-calving cow therefore tends to have a steep rise to a high peak yield and a sudden decline, compared with the autumn-calver, who makes less use of grazed grass (Fig. 4.1). In autumn, the quantity and quality of the grass is usually low, even though an autumn flush that follows late summer rainfall may suggest that there is young, high quality pasture available. It is certainly of less value than herbage produced in spring, even though ME values are often similar, probably because protein degradability in the rumen is high, the moisture contents are low and much of the herbage is rejected because of its proximity

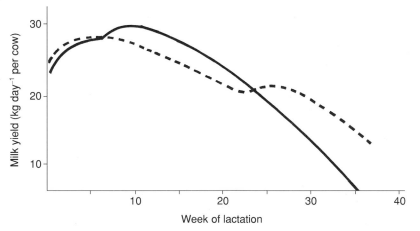

Fig. 4.1. Typical lactation curves for autumn-calving (- - -) and spring-calving (——) dairy cows.

to faecal deposits. In addition, the physical composition of autumn grass is different to grass produced earlier in the year, since there a dense mat of dead grass leaves and stems at the base of the sward that has escaped defoliation, which cattle are reluctant to consume. Thus, lush leafy grass grown in autumn will not support high milk yields unless it is supplemented with forage and concentrate feeds.

Stocking Density

The farmer's objective in determining the stocking density of cattle is to provide adequate herbage of high quality. This requires skill in ensuring that the grass is kept young and has a high leaf content, but still is present in sufficient quantity. Some cattle, particularly high-yielding dairy cows, never achieve high intakes from pasture because their rate of intake is too low.

As stocking density increases from low to medium levels, so the milk production per cow declines but milk production per unit area increases. At high stocking density the milk production per unit area may decline as well as the production per cow, since an excessively large proportion of the intake is used to maintain the cow. Also, herbage growth may decline as the leaf area index of the grass is reduced by excessive defoliation.

The optimum stocking density should provide sufficient herbage to each cow to allow her to produce 90% of her potential milk production, the remainder coming from supplements. For a good quality sward that is well supplied with fertilizer in warm temperate conditions, this is likely to be 6–7 cows ha^{-1}. Such conditions are usually present up until the first cut of silage in about late May in Britain. The stocking rate can be relaxed after this to perhaps 4 cows ha^{-1} by introducing silage aftermaths and perhaps relaxed again to 3 cows ha^{-1} after a second cut of silage at the end of July.

At its best, fresh herbage is of considerably greater nutritional value than conserved forages, since the latter can only be efficiently harvested when the grass is tall and relatively mature. However high-yielding dairy cows may have difficulty harvesting sufficient material in the time available if the grass is too short, and they will then quickly lose weight. Cattle are reluctant to graze for more than one-half of the day, and most lactating cows spend about 8–10 h in this activity. As grass has a high fibre content, they need to spend 6–9 h daily ruminating, most of which is when they are lying down at night. Cattle prefer not to graze during darkness, which limits the time available for grazing. If herbage availability declines, cows increase their grazing time and biting rate in an attempt to compensate for small bites, but cows on very short pasture may stop grazing if they are fed supplements.

Intakes of DM are usually less when cows graze herbage than when they are fed conserved feeds, because of the low availability of pasture. However, in spring, the high digestibility of grazed herbage encourages high milk yields, at least whilst the cows still have fat reserves to support their energy requirements. Later, as herbage declines in quality and quantity, and energy reserves become exhausted, the milk yields can decline rapidly. In late summer, it is extremely difficult for cows to produce high milk yields from pasture alone.

In intensive grassland systems in the UK, where predictable milk yields are important for quota management, farmers tend to understock their pasture to ensure that grass is still available if grass growth declines. Fortunately in dry conditions, the high DM content of the grass encourages high DM intakes. However, in wet conditions, if the water content of the grass, including surface moisture, exceeds about 80%, then intake can fall rapidly. This is because of the difficulty cows have chewing wet grass and the reduced saliva produced to get the bolus to a suitably moist form to swallow. In Britain, farmers in the east of the country cannot grow as much grass as those in the west, because of the low rainfall and the reduced influence of the Gulf Stream, making the winters longer. However, in the relatively dry weather, the cattle make better use of the herbage that is available. In the wetter west, more grass is wasted by the cattle through trampling losses and a failure to match growth with require-ments, with the result that there is little difference between farmers in the dry east and wet west of the country in the amount of utilized herbage from each area. Good farmers are conscious of the fact that they are dealing with a dynamic process – grass is constantly growing, in summer at least, and if it is not harvested within a few weeks, the leaves senesce and are of no use to the grazing cattle. Senesced leaves are not completely wasted since the plant retranslocates much of the valuable mineral content to growing leaves and the residue will eventually be absorbed into the humus layer of the soil to support further growth. However, greater output per unit area is undoubtedly achieved with high stocking densities, where a greater proportion of the grass grown is actually harvested.

Measuring Grassland Production

Grassland production and utilization for a farm can be calculated as the utilized metabolizable energy (UME). This is an indirect measure determined from the cows' ME output, usually over 1 year, in relation to the land area that they utilize. First the ME output of the cows is calculated from tabulated values of ME requirements for milk production, weight change and body main-tenance (see Chapter 3). The ME contributions from feeds that have been imported onto the farm are then deducted. The area utilized by the cows has to be estimated, which can be difficult if other stock have used the land. Sheep may have been grazed on the farm during winter or some of the conserved forage used for other stock. The product of mean cow ME output per year from herbage (grazed and conserved, in GJ), the number of cows and the land area, gives the UME (GJ ha^{-1}).

UME = ME output per cow per year from herbage × number of cows × land area.

Typically research farms are able to achieve about 120 GJ ha^{-1} on small areas of land, good dairy farmers about 80 GJ ha^{-1} over the whole farm and average farmers about 50 GJ ha^{-1}. Generally UMEs are closely correlated with farm profitability. Although it is a useful measure of grassland utilization, the UME should not be the only measure of efficiency – an overstocked farm may have a high UME but a large proportion may be used for cow maintenance at the expense of milk yield. Also, some farmers achieve high UMEs by applying a lot of fertilizers, which does not necessarily mean that their grassland utilization is any more efficient than a farmer using less fertilizer. Thus, although less grass may actually be grown if stocking rates are high, more of the grass that is grown is utilized. The UME can be combined with an estimate of grass growth, as predicted from the site class and level of fertilizer application, to give an estimate of utilization efficiency.

Grass height is the most practical measure of herbage availability, although the ability of any grass sward to support high intakes by the cattle depends not just on the grass height but also its density. These are the two main features of the sward that affect intake. Others include the proportion of leaf relative to stem, the nature of the leaves, hairy or smooth, and the degree of soiling of the leaves with faeces or soil. Of importance to the growth of grass is the leaf area index, or leaf area per unit soil area, which should be approximately 5 for optimum growth. This, however, is not easy for farmers to measure, whereas height can be used to determine whether the optimum quantity of grass is available to support adequate intakes, and whether the grass is tall enough to be growing efficiently. The height can be measured with a sward stick, with a stick and sliding sleeve placed on the ground. The sleeve has a needle on it which is lowered until it touches the tallest tillers, with the height then being read off the calibrated stick (Barthram, 1986). Alternatively a plate of about 30 cm^2 may replace the needle and again the height is read off the calibrated stick. This 'rising plate meter' reduces the measured heights by about 1 cm because of the compression of the grass by the weight of the plate. It is unsuitable for measuring grass heights of less than 5 cm as the grass cannot support the weight of the plate.

As the grazing season progresses, the sward develops into a mosaic of short and tall grass, with the areas of tall grass in circles around faecal deposits. Cows mostly graze the short grass areas, but when these become too short, they will consume the grass around the faecal deposits.

Grazing Systems

The dynamic nature of grass growth makes it preferable to be flexible in the movement of cows around the pastureland. It helps if the milking unit is in the middle of the grazing block, otherwise cows and the herdsperson may have to walk long distances to the milking parlour. Usually, the herdsperson prefers to have the cows close to the milking unit overnight so that they do not have far to walk early in the morning to collect them for the first milking, and this can lead to a transfer of soil fertility to these fields. Within a field, the land by the boundary hedges is often the most fertile, as cattle shelter there, especially overnight, and return more excreta to the land.

The simplest grazing systems keep cows on the same pasture all the time, although extra land may be added after silage cuts to compensate for reduced growth as the grazing season progresses (Fig. 4.2). Such systems are called continuous grazing or set stocking. Alternatively, a farmer may rotate the cows around the grazing area, which may be split up into fields of several hectares each with cows in for a few days at a time, or into 15–25 paddocks where cows would normally only spend 1 day at a time. In paddock grazing, it there-fore takes 2–3 weeks for a complete rotation. Cows should be removed from a paddock when pasture height is about 8–10 cm if intake is to be maximized (Fig. 4.3). This critical grass height, at which DM intake begins to decline, is usually 1–2 cm lower for continuously grazed swards because the herbage density is greater. The net herbage growth (herbage growth – senescence) is maximized in a temperate sward at 3–5 cm, but there is only a slight reduction if the sward height is 5–8 cm. If herbage is maintained at 9 cm or above, the grass develops many stem internodes and becomes highly lignified and of low digestibility. The amount of leaf present is likely to be no greater than that present in a sward maintained at 3 cm. If the herbage height is reduced below 3 cm there is a considerable reduction in growth rate. Also when the herbage is very short, cattle may uproot tillers when grazing and the plant is often not

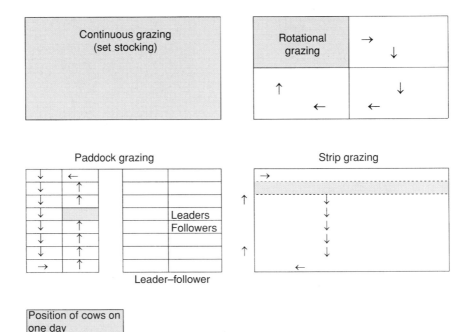

Fig. 4.2. Diagrammatic representations of the continuous, rotational, paddock and strip-grazing systems.

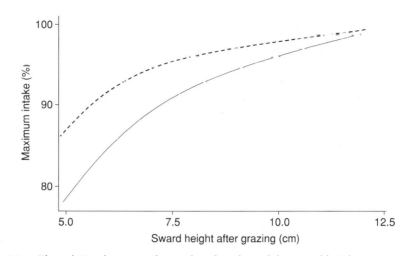

Fig. 4.3. The relation between the intake of cattle and the sward height post grazing for rotational (——) or continuous (- - - -) grazing systems.

consumed but left on the pasture. When the herbage is very short the cattle have little opportunity for selection; however, when the herbage is tall, cattle select leaf in favour of stem and green material in favour of yellow/brown grass. Such herbage contains more nitrogen, phosphorus and energy. The only occasion when cattle will avoid lush, green herbage is when they graze a high clover sward, when they sometimes make a deliberate selection of the more fibrous grass, which contains less energy and protein than the clover. This is presumably to avoid the risk of ruminal tympany.

One of the main advantages of rotational grazing is its flexibility, providing the opportunity to shut up fields or paddocks for conservation if they are not required for grazing. Another advantage is that herbage availability can be assessed more easily with rotational grazing. Scientific experiments have demonstrated that there is no inherent advantage to either rotational or continuous grazing. In wet conditions, some paddocks may suffer severe poaching damage when hooves destroy the sward structure. In continuous grazing systems, the damage may be less severe but more widespread. A disadvantage of rotational grazing is the requirement for a large amount of fencing, now usually electric, and many water troughs and gates. It is, however, usually quicker to get cows in for milking from a small paddock than several fields. Farmers should resist the temptation to round up the cows for milking with a dog or motor vehicle, as they tend to be rushed down the track, leading to more lameness, and poor milk let-down in the parlour.

Rotational grazing requires more cow tracks. All cow tracks to the pasture should preferably be concrete or some other durable material, because in stony soils, particularly those with flints, the stones can cause a high incidence of punctured soles, white line separation and solar abscesses (see Chapter 6). This is particularly true in wet conditions when a muddy track provides an unsuitable surface for the cows to walk on.

Another way of rotating cows around a field is to divide it into strips using an electric fence. Usually just one fence is moved across the field and cows have access to previously grazed strips and a water trough, but sometimes a back fence is included to allow grass in the previous strips to regrow before the cows have consumed all the strips in the field. New strips are offered to the cows at least once daily and most farmers like to move the fences so that the cows have a fresh strip after each milking. Strip-grazing allows effective rationing of the grass to the cows, and may limit the treading damage to the pasture as the cows spend only limited time on each day's ration, spending the rest of their time on bare pasture.

A system that helps to overcome the problems of inadequate grass for high-yielding cows is the leader–follower system. Here a herd is divided into high- and low-yielding groups and the former graze paddocks or fields first before the low yielders. In practice, unless the followers are very low yielding or dry, the advantage in yield to the high yielders is offset by a similar reduction in yield in the low yielders.

Rotational grazing is better than continuous grazing for mixed grass and clover swards, as clover likes a period to recover after heavy grazing. In continuous grazing, cows can persistently select the area of high clover content, which becomes depleted. Rotational grazing also suits an extended grazing system in winter, allowing the cows access to small areas of herbage to (i) augment their winter ration, and (ii) give them some exercise and better conditions underfoot. However, if they are allowed access to a large area they may damage the sward structure with their hooves and cause poaching damage. Free drainage and light soils are essential. In mild temperate areas it is quite common for cows to be given a few hours access to pastureland each day in fine weather. In areas where there is a risk of frost damage, tall leafy grass should not be kept for grazing as it may be killed off by the frost (winter kill), and a large dead matter content in spring will retard new growth. In such circumstances, it is better to use sheep to graze the pasture down to 3–4 cm over the winter, ensuring that they are removed before the start of spring so that new growth is removed for the cows. Mature herbage can be saved for winter grazing (foggage or standing hay) but is not of much nutritional value for the cows. Extended grazing can reduce the need for concentrate supplements and improve the cows' performance by supplementing silage with leafy herbage.

A final system for utilizing fresh herbage is cutting the herbage on a regular, usually daily, basis and carrying it to the cattle, sometimes called zero-grazing. In temperate climates, the cattle are housed, but in hotter climates they may simply be provided with a roof for shelter. The system is mechanically and labour-intensive, but it makes good utilization of the herbage that is grown since it ensures a complete cut and minimal wastage. Some damage to the land is possible in wet conditions. It does not require the land to be fenced or water to be provided and may be favoured if cows have to voluntarily visit an automatic milking plant regularly or there is a risk of cattle being stolen or attacked if they are out in the fields. The impact of permanent housing systems on cattle health is discussed in Chapter 6.

Youngstock Grazing

Youngstock usually do not receive the accurately rationed grass that lactating cows require. Often they are set stocked at a low rate on a distant part of the farm, as they do not need to be regularly brought to the farm as milking cows do. There is some advantage of rotational grazing in the control of gastrointestinal worms. Often, however, the life cycle of the worms is longer than the rotation and farmers cannot be sure that there are no infective larvae after one rotation. Cattle in their first grazing season are usually dosed with anthelmintic and moved to pasture that has not been grazed that season. A leader–follower system, where animals in their first grazing season lead the rotation and those in their second season follow, offers some protection to the calves against gastrointestinal worms (Table 4.1).

If rotational grazing requires more labour than is available on the farm, care should be taken to move the cattle to clean grazing at least once a year. The reduction in grass growth as the summer progresses can be accommodated if the cattle graze one-third of the grazing area for the first third of the grazing season, up to a first cut of silage, then two-thirds of the area for the second third of the season, at the end of which a second silage cut is taken and finally the whole area for the last third of the season. Some infective larvae can overwinter, since the life cycle is extended in cool conditions, and the only way to be sure that pasture does not have infective larvae is by alternating cattle and sheep grazing annually, or grazing cattle on land that has only been used for conservation in the previous year. Gastrointestinal worms are species specific, but the liver fluke (*Fasciola hepatica*) can infect both cattle and sheep. The modern method of protecting cattle from gastrointestinal worms is to use anthelmintic-releasing intra-ruminal boluses, but this may lead to older cattle and even lactating cows being susceptible to infection, as immunity does not accumulate during the rearing period.

The growth of cattle may be checked if their diet is suddenly altered, as a result of the time required for the rumen microflora to adapt. After a sudden change, cattle often do not gain weight for 2 weeks or more. The changes in

Table 4.1. The effects of providing an anthelmintic to calves grazed in a leader–follower system or in separate rotations (from Leaver (1970), with permission).

	Leader–follower		Separate rotations	
Anthelmintic	Yes	No	Yes	No
Calf growth rate[a] (kg day^{-1})	0.79	0.79	0.63	0.48

[a]No difference was observed in heifer growth rate.

weight of an animal can be deceptive, either because it is weighed at different times of the day, or because gut fill varies with the digestibility of feeds. There can be a weight reduction of 20 kg or more when cattle are transferred from a diet of conserved feeds to lush pasture, as the high digestibility of the latter increases the rate of passage through the gastrointestinal tract and reduces the weight of the contents of the tract. It is particularly important to buffer the transition from a conserved forage diet to grazed herbage by continuing to offer conserved feeds at pasture for at least 2 weeks. These can be offered in racks or round bale feeders in the field, preferably on a hard standing area if the field is susceptible to poaching damage. Conserved feed can also be used to buffer changes in herbage availability or quality.

Pasture Supplements

Grazing cows giving high yields usually need concentrate supplements. The level of milk production that can be supported from pasture alone depends on the herbage availability and quality and the cow's body reserves. On temperate pasture, cows are rarely able to give more than 25 l day^{-1} milk from pasture alone for more than a few weeks. Supplements provided to grazing cows in temperate countries should be essentially energy supplements, although very high-yielding cows are likely to increase milk yield in response to protein that bypasses the rumen. Concentrates may be provided to grazing cows according to pasture height and milk yield. The critical grass height below which the cows' intake begins to fall is lower in autumn than spring because much of the grass, often up to 30–40%, is rejected because it is close to faeces. Grass is initially rejected around faeces because the cows avoid grazing where they can smell their own faeces – a natural anti-parasite strategy. Later, a crust forms on the surface of the faeces, particularly if the weather is warm and sunny, but the herbage is still rejected, even after the faecal deposit has become partly decomposed and is of no risk to the cattle, because the nutritional value has declined. Faecal deposits are degraded by insects, bacteria and worms in a few months in warm conditions. In winter, however, it can take 6 months or more for them to break down because of reduced activity of the organisms responsible for the digestion of the material.

Pasture supplements are important for grazing cows to:

1. Increase intake when herbage is in short supply.
2. Ease the transition to or from a diet of conserved feed at the beginning and end of the grazing season.
3. Increase the intake of high-yielding cows that cannot consume enough grass to sustain a high milk yield.
4. Increase the fat content of milk, in the case of forage supplements.
5. Maintain nutrient intake when inclement weather reduces the intake of grazed herbage.

The maximum milk yield that can be sustained on grass alone (about 25 l day^{-1}) assumes that the herbage is high quality (at least 12 MJ ME kg^{-1} DM). The intake of herbage may be insufficient to support high levels of production from a dairy cow, because of low herbage mass/bite, inadequate grazing time (especially when photoperiod declines in the autumn), a high herbage moisture content, a high dead matter content and a high faecal contamination. If adequate pasture is available, the reduction in herbage DM intake for each unit of supplementary concentrate fed may approach or even exceed one. The substitution rate can be calculated as:

$$\frac{\text{reduction in DM intake of basal feed (kg)}}{\text{DM intake of supplementary feed (kg)}}$$

Usually a concentrate supplement is more rapidly digested than the basal herbage and so the cow increases her intake with each additional portion of supplementary feed, i.e. the substitution rate < 1. If, however, forage supplements are offered with pasture, the substitution rate may be > 1 if they are digested more slowly and total DM intake is reduced (Phillips, 1988). Why do cows eat such supplements if better quality grass is available? Firstly, conserved feeds are easier to consume rapidly and some cows prefer to stand by a silage feeder consuming their feed rather than spending their time grazing. Thus, the cow does not necessarily always optimize energy intake but may also takes account of the ease with which feeds can be consumed. This could optimize energy retention, rather than intake, if the energy expenditure in food procurement is taken into account. Secondly, young leafy grass may lead to a rather unstable rumen fermentation, with low rumen pH, low milk fat contents and even bloat. Consuming supplementary forages will stabilize rumen conditions and increase milk fat production, even if milk yield and protein content are reduced. Thirdly, cows may maintain a varied diet deliberately so that rumen microflora are prepared for future changes in diet.

The advantages of providing forage supplements is twofold. First, they are less expensive than purchased concentrates per unit of energy, and secondly, cows are usually reluctant to consume large quantities of forage supplements in preference to grass, unless the latter is severely restricted. It may be difficult for the farmer to decide when the grass is too short for the cows. Measuring herbage height is time consuming and can be misleading because other factors, such as the density of the grass and the species, also affect intake. Offering the cows conserved forage for a short period each day, perhaps for 1 h after morning milking, allows the cow to make the decision as to whether she is getting enough feed from the available pasture. The supplement should preferably be offered at a time when the cow would not normally be grazing intensively, e.g. not just after the afternoon milking, which is normally the time when the longest grazing bout begins. After the morning milking, however, cows often lie down to ruminate after their early morning grazing bout. If feed is offered at this time, good quality hay is best as it can be left from day to day without spoiling. Opening silage clamps in summer can lead to serious spoilage at the surface unless the clamp is long and narrow and silage removed by a block cutter. If farmers are keen to use this system of 'buffer feeding' on a regular basis, it may be worth investing in clamps that can be filled at either end, allowing feeding at one end and filling at the other. Cows will not usually consume the conserved forage for more than about 30 min to 1 h, and should be returned to their pasture promptly. If pasture is in short supply, cows can be kept indoors overnight, particularly if they are being milked three times a day. This will reduce grazing time and may cause hoof problems if cows are housed in wet passageways and then turned out with soft hooves onto stony land. This could also be a problem with extending the grazing for part of the day in winter. Maize silage is particularly suitable for overnight feeding of grazing cows, as the high-energy, low-protein content of the silage complements medium-energy, high-protein grass well. Forages can also be fed continuously in the grazing field, but there is likely to be more feed wastage and poaching damage around the feeder.

The main advantage of routine buffer feeding is an increase in the predictability of the cow's performance. This is important for the management of a farm's quota allocation. A cow's intake of grazed grass is particularly dependent on the weather conditions, as well as factors more under the control of the farmer such as herbage availability. The intake of forage supplements is usually high in the autumn when the apparent energy value of the grazed grass in low.

Concentrate supplements may be better than forage supplements because of a lower substitution rate, a higher rumen undegraded protein content and less waste. Usually concentrates are more expensive than silage per energy unit, which is in turn more expensive than grazed grass. Maximizing yields of milk from grazed grass has been, and will continue to be, a key objective for profitable milk production. With low-yielding cows, if adequate herbage is available, the substitution rates of concentrates for grazed herbage may approach 1, i.e. for every 1 kg DM of concentrate supplement offered, the cow reduces her herbage DM intake by close to 1 kg. This is clearly unprofitable use of concentrate supplements. However, if cows are giving high yields over several weeks (>25 l day^{-1}), they will need a supplement of least 2–4 kg day^{-1} of concentrate to achieve an adequate energy intake. Such cows can also benefit from extra rumen-undegraded protein in the concentrate. As concentrates are generally consumed in preference to herbage, it is more difficult for a farmer to ration concentrate supplements to grazing cows than forage supplements, which tend to be eaten mainly when the intakes of grazed herbage are inadequate.

A high magnesium concentrate may be fed at a low rate (c. 1 kg day^{-1} per cow) for 1–2 months after turnout as a means of preventing hypomagnesaemia (see Chapter 3).

High Clover Swards

Cows on a ryegrass and white clover sward will produce greater milk yields or cattle growth rates than those on a pure ryegrass sward, caused mainly by the higher digestibility and nitrogen content of white clover compared with the ryegrass (Table 4.2). White clover has less hemicellulose and cellulose, although it has slightly more lignin. Often, the white clover has a greater mineral content than ryegrass, particularly magnesium, calcium, iron and cobalt. Even when compared at similar digestibilities, clover supports higher milk production and growth rates than grass. Cattle will normally select areas of the sward with high clover contents, which may lead to its depletion in the sward (Phillips and James, 1998).

Mixed grass and clover swards should aim to have at least 30% clover to achieve the equivalent of 200 kg N ha^{-1} fixed. Up to 280 kg N ha^{-1} is possible, but on average only one-half of that is achieved in the UK. Stocking densities on high clover swards are therefore likely to be only two-thirds of that on a highly fertilized sward containing only grass, since the grass responds to up to about 400 kg N ha^{-1}. White clover (*Trifolium repens*) is more persistent and disease resistant than red clover (*Trifolium pratense*). Maintaining this much clover can be difficult, but the clover content should not be greater than 50% if bloat is to be avoided. High clover contents often occur soon after seeding if the establishment is good, but the establishment of clover is unpredictable. In hill pastures, it requires seed inoculation with the appropriate *Rhizobium* species. Usually 3–4 kg clover seed is included in the seed mix per hectare. Low soil pH will hinder establishment and growth, as will inadequate calcium, potassium and phosphorus status of the soil. Large leaf varieties are used for lowland swards and small leaf, wild white varieties for marginal areas. As the sward matures, the nitrogen fixed by the rhizobia associated with the clover in root nodules promotes grass growth. However, since grass has a more erect profile than clover, the light reaching the clover plants is often reduced as grass growth is increased, and the clover content starts to decline.

An additional problem is that white clover starts growing later in spring, as it requires 9°C to commence growth, compared with 6°C for grass. Thus, it may be shaded out by fast growing grasses in early season. As a result of this, the clover content may start at 20% of the sward in spring and increase to 40% by mid-season. It is more susceptible to drought than most grasses. It is unwise

Table 4.2. Typical composition (g kg^{-1} DM) of clover and perennial ryegrass.

	Perennial ryegrass	White clover
Nitrogen	28	44
Cell wall	427	216
Cellulose	240	173
Lignin	27	38
Hemicellulose	161	8

to use nitrogen fertilizer or large amounts of farmyard manure or slurry on high clover swards, because of the promotion of grass growth. However, it is possible to provide an application of 50 kg N ha^{-1} in spring to achieve additional grass growth before the clover starts growing. Such a sward must be grazed well to prevent the grass shading the clover. It may even be beneficial for the sward to have vigorous grass growth in spring to prevent weeds becoming established. Weed control is difficult in high clover swards because many broad-leaved herbicides cannot be used as they kill the clover.

Sheep can deplete the clover content of a sward in winter if they are brought onto a dairy farm for winter keep and graze the sward very low. They also select clover plants in the sward in a way that cattle, with their broad dental arcade, cannot do. A period of rest is required for white clover to recover; otherwise, sustained defoliation will deplete reserves and reduce the clover growing points. For this reason, rotational grazing is preferred. The difficulties in maintaining a high clover content in a sward mean that clover growth is less predictable than grass and it is therefore less persistent.

Bloat

Bloat can be observed as an acute swelling between the last rib and the hip on the left side of the cow, from behind. The cow is restless, finding lying uncomfortable and may eventually die of heart failure or suffocation as a result of inhaling rumen contents.

Pasture bloat is caused by a stable foam in the rumen caused by the rapid digestion of legumes, in particular, although young leafy grass that has recently received nitrogen fertilizer can also cause bloat. Lucerne is the most likely of all legumes to cause bloat, with cows sometimes dying within a few hours of entering a field for grazing. Some legumes have developed a chemical, tannin, which reduces the speed of protein digestion and probably discourages animals from grazing it. They are present in sufficient quantities in birdsfoot trefoil to prevent the production of a stable foam, and the content in white clover increases sufficiently at flowering to make it safe to graze. If a mixed grass and clover sward has enough clover to cause bloat (probably more than 50% of the herbage by mass), it should not be grazed for long periods, but should either be conserved if there is sufficient mass, or it should be rested for a few weeks until the clover inflorescences appear, after which it can be grazed or conserved.

Cows are most likely to become bloated in the late evening after a day's grazing and also after a wet period when they avidly graze to make up for lost time. Wet grass reduces saliva production, and the saliva contains a mucin that disperses foam in the rumen. Herbage that has been frozen is particularly likely to cause bloat as the rupture of plant cell walls releases the solutes that contain much potassium. Potassium-rich feeds, such as molasses, are well known for causing bloat, whereas grasses rich in sodium appear to be less likely to cause it. The precise mechanism has not yet been determined but may relate to the stimulation of saliva production by sodium-rich feeds, and the foam-dispersing properties of the salivary mucin

Forage supplements will usually slow down the rate of digestion and reduce bloat, but if there is adequate herbage, grazing supplements may not be eaten by some cows in sufficient quantities, particularly if they are based on straw or other low quality forages. Mineral oils also help to disperse the foam and can be added to a concentrate feed or sprayed onto the pasture or the cows' flanks, to be licked off as needed. Linseed oil is often used. A proprietary product, poloxalene, also breaks up the foam and can be used as a drench for clinical cases or included in feed blocks as a preventative measure. Often simply walking the cow from the field to the farm steading to receive medication will alleviate the swelling. It is important to keep a bloated cow on her feet if possible, as death can follow soon after recumbency.

There is undoubtedly a genetic component in the susceptibility of cattle to bloat. There are reported breed differences in susceptibility – Jersey cows are particularly prone to the disorder. Cows can get used to feeds that are liable to make them bloat, this may be by altering their behaviour to spread their meals out more evenly over the day. Lactating cows are particularly susceptible because of their high intakes.

Pasture bloat remains a serious problem for farmers in countries like New Zealand, where the cattle rely on pasture with little or no fertilizer applied and a high legume content. In Europe, the greater emphasis on controlling nitrogen emissions may encourage farmers to use high clover swards for their cattle, potentially leading to more serious problems with bloat.

Fertilizer Application

All grassland needs some return of nutrients if the land is to remain fertile, to replace the nutrients that are removed by the animals and are lost from the soil by leaching. However, over-application of fertilizer, particularly potassium and nitrogen, is wasteful and can lead to pollution of ground water. Less nitrogen is leached from high clover swards, compared with highly fertilized grass swards.

Nitrogen is the first limiting nutrient on most sites and is applied in the largest quantities. In Britain the annual herbage DM yields range from 8 t ha^{-1} for a poor quality site with 300 kg N ha^{-1} applied to 13 t for a first class site with 450 kg N ha^{-1} applied. Most grassland farms apply less nitrogen than this, the average on dairy farms being about 150 kg N ha^{-1}. Considerably less than this was applied up until the 1970s in the UK. High profitability of the dairy industry in the 1980s increased nitrogen use significantly, but at the end of the 20th century, fertilizer applications reverted to lower levels in response to economic and ecological pressures. The time of application of the first nitrogen of the season is critical to the pattern of grass growth. Too early and it may be leached through the soil to groundwater without having been used by the grass plant. Too late and spring growth will be retarded.

ιe optimum time of first application is best estimated by summing the mean
αily temperature from 1 January, and in Britain this should be about 200°C.
Most nitrogen should be applied in spring and early summer. At this time it is
unlikely to be leached out of the soil as the grass crop is growing fast and
taking up nitrogen, and the soil moisture deficit usually is increasing as a result
of the high evapotranspiration. The penalty for early application of nitrogen is
that a large proportion of the annual production of herbage is grown early in
the season, which may be far more than the requirements of the cattle. This
gives an opportunity for the farm to conserve one or two cuts of silage early in
the season to provide the winter forage. Some will take a third or even a fourth
cut later in the season, perhaps up to late August. Frequent cutting will
produce high yields of good quality forage.

Although most of the nitrogen is usually applied in spring and early
summer, phosphorus is less likely to leach and can be applied at any time in
the growing season. Often it is applied at the beginning of the season so that it
is available if needed at any time in the season. Potassium fertilizer should not
be applied in the spring as it is associated with reduced magnesium availability
to cattle, and may trigger hypomagnesaemia. It is needed less for grazed crops
than silage, because of the return of potassium in excreta in the grazed sward.
The low magnesium content of rapidly growing grass swards as well as the
low availability of the magnesium often present difficulties on farms where the
cows are susceptible to hypomagnesaemia. The advantages of supplying
sodium and sulphur fertilizers have been discussed in Chapter 3.

In summary, fertilizers are essential for maintaining the fertility of the land
when nutrients are removed by the grazing animal. The grazing animal returns
many of the nutrients that are consumed but some may be leached from the
soil as they are returned at high concentrations. This is particularly true for
urine, rather than faeces. If the sward is conserved rather than grazed, more
fertilizer, particularly potassium, will have to be applied to the land. This may
be as artificial fertilizer or as livestock manures. Particular care is required in
the return of livestock manures to the land, to avoid high nutrient applications
at times when the rainfall is likely to be high and the soil moisture deficit low.
Smearing of the grass can also occur in spring, which can reduce grass growth
and the intake of the grass by the cattle.

References

Barthram, G.T. (1986) Experimental techniques: the HFRO sward stick. Biennial Report,
    Hill Farming Research Organisation, 1984–5, pp. 29–30.

Krohn, C.C. and Rasmussen, M.D. (1992) Malkekoer under ekstremt forskellige
    productionsbetingelser: ydelse, tilvaekst, reproduktion, sundhed og holdbarhed.
    [Dairy cows under extreme conditions: production, reproduction, health and
    stayability.] Landbrugsministeriet Statens Husdyrbrugsforsog (Danish Institute of
    Animal Science) Report No 70. DIAS, Foulum.

Leaver, J.D. (1970) *Journal of Agricultural Science (Cambridge)* 75, 265–272.

Phillips, C.J.C. (1988) Review article: the use of conserved forage as a buffer feed for grazing dairy cows. *Grass and Forage Science* 43, 215–230.

Phillips, C.J.C. and James, N.L. (1998) The effects of including white clover in perennial ryegrass swards and the height of mixed swards on the milk production, sward selection and ingestive behaviour of dairy cows. *Animal Science* 67, 195–202.

Further Reading

Cherney, J.H. (ed.) (1998) *Grass for Dairy Cattle.* CAB International, Wallingford.

Frame, J. (1992) *Improved Grassland Management.* Farming Press, Ipswich.

Hodgson, J. (1990) *Grazing Management: Science into Practice.* Longmans' Handbooks in Agriculture Series, London.

Hodgson, J. and Illius, A.W. (eds) (1996) *The Ecology and Management of Grazing Systems.* CAB International, Wallingford.

Hopkins, A. (ed.) (1989) *Grass, its Production and Utilization*, 3rd edn. Blackwell Scientific Publications, Oxford.

Mannetje, L.t' and Jones, R.M. (eds) (2000) *Field and Laboratory Methods for Grassland and Animal Production Research.* CABI Publishing, Wallingford.

Wilkinson, J.M. (1984) *Milk and Meat from Grass.* Granada Technical Books, London.

Breeding and Reproduction 5

Introduction

Cattle are polygynous animals, which has resulted in significant sexual dimorphism. The bull is larger and stronger, particularly in the neck, shoulder and size of horns, which contribute to his ability to fight for access to females. In feral cattle herds, males live a quite solitary existence, leaving the herd once they reach sexual maturity. Older males dominate the younger bachelor males and have priority of access to females. During oestrus, the females indicate to distant males that there are receptive animals in the matriarchal group by mounting each other. Males also mount each other in intensive husbandry conditions, but this is probably redirected aggression rather than any evidence of sexual motivation.

Reproduction in feral cattle is more synchronized than in farmed cattle, so that most cows give birth in spring, and peak lactation coincides with the period of maximum food availability. Modern, domesticated cattle still show some seasonality of reproduction as a result. Neonatal development is rapid, which is usually the case for prey animals. The precocious calves stand rapidly and usually suckle within 6 h of birth, after which the permeability of the gut to the passive transfer of immunity from the cows rapidly diminishes.

With domestication came changing requirements for cattle, as they were destined to become man's main meat and milk provider. The new environment in which cattle were kept post-domestication provided new challenges for cattle breeders, as seasonality of calving became a disadvantage, and improved feeding potential favoured those with high milk production and rapid growth rates. Originally opportunities for selection would have been limited, but gradually breed improvement has taken place and increasingly effective techniques to bring about the required changes have developed.

The Start of Systematic Breed Improvement – Bakewell's Influence

As far as we can tell, the form, behaviour and productivity of cattle has changed considerably since they were first domesticated 8000–10,000 years ago. Over time, natural selection has gradually been complemented and, to some extent, replaced by artificial selection. During the early phase of domestication, primitive farmers appear to have selected for the following characteristics:

1. Lack of aggression/docile temperament.
2. Short flight distance in reaction to human presence.
3. Small and manageable size.
4. Ability to adapt to an unnatural environment.
5. Willingness to eat non-conventional feeds.
6. Overt sexual behaviour in the female.

Opportunities for selection must have been limited in the early days of domestication, particularly when village cattle were communally grazed. Selection was probably based mainly on the restricted use of a limited number of bulls. Later, during the industrial revolution, the pace of breed improvement was increased, in an attempt to meet greater demand for cattle products in the newly industrialized countries.

Robert Bakewell (1725–1795) was one of the first English farmers to attempt to improve radically the quality of cattle, and he was unique in his era for two characteristics. Firstly, at a time when most farmers practised cross-breeding, he selected a breed of cattle that he believed would respond well to selection, the Longhorn, and used inbreeding (selecting within a breed) to achieve genetic improvement. Secondly, while most farmers in Britain were using cattle for both milk and meat production, he developed his selected breed exclusively for meat production. He selected in particular for the ability to fatten quickly, and to develop subcutaneous fat deposits in the hindquarters. Eighteenth-century labourers needed to consume more energy than many of today's workforce, and they preferred fatter meat than we consume nowadays. Also, surplus fat was valuable for tallow used in candles.

Bakewell's influence on livestock breeding spanned the early years of the industrial revolution, and indeed, he was a key component of the agricultural revolution that started in the mid-18th century. The early introduction of land enclosures in England gave farmers better control over cattle breeding, and thereby gave them more scope for breed improvement. Bakewell's legacy was perhaps not so much the improved Longhorn breed, as this proved to be of limited value in England, but the way in which he managed to change this and other breeds through a process of scientific research. He kept meticulous records, but as is increasingly common with industrially sponsored agricultural research nowadays, he did not divulge these to others. The end result of his labours was sufficient evidence in itself – an animal that was clearly more useful for meat production than the original Longhorns that had previously been mainly used for draft purposes in the days before mechanized tillage.

However, Britain wanted a dual-purpose animal, for milk and meat production, and for this reason, the improved Longhorn was less successful than the Shorthorn that was developed at the same time for both milk and meat production.

Bakewell was the first in a long line of pioneer breeders in Great Britain who developed breeds for a variety of purposes. In the 18th and 19th centuries, the variety of British cattle that had been developed was to be particularly useful during the expansion of the British Empire, when cattle with different characteristics, such as heat resistance, were needed to feed the expanding populations at home and in the colonies. The British population increased in number from 7 million in 1760 to 31 million in 1881, and the greater affluence associated with industrial development increased the demand for beef.

Bakewell was also ahead of his time in the way in which he managed his cattle. He placed great importance on fertilizing his pastures with manure, leading to increased production per unit area. This was probably forced on him as a result of his farm being quite small. He kept his cattle in individual stalls in winter, which reduced poaching damage to his pastures, and bred his Longhorn cattle with ingrowing horns (bonnet style) to enable them to be stocked at a high rate. The horn of cattle in those days had many uses, such as the manufacture of combs, buttons, knife and whip handles and a cheap lantern glass when prepared in thin section, and was another valuable attribute of the Longhorn breed.

More recently, the reluctance of British cattle producers to relinquish their dual-purpose animals, in common with producers in many parts of Europe, has stood in stark contrast to the former colonies where single purpose cattle prevail. These are usually more efficient at producing either meat or milk than the dual-purpose breeds. The recent reduction in demand for beef, as a result initially of concerns about the high levels of consumption of saturated fat and later because of the risk of acquiring Creutzfeldt–Jacob disease from meat infected with bovine spongiform encephalopathy (BSE), is now making cattle farmers concentrate on milk production. Previously the income from calves for meat production made a greater contribution to their total income, so an integrated industry evolved.

Modern Cattle Breeding

The first rapid increase in the development of cattle breeds during the agricultural revolution of 1750–1880 was followed by a period of consolidation of the breeds that had been developed. With two world wars in the first half of the 20th century, agriculture was in a depression by 1950. In the latter half of the century new technologies were implemented to meet an increased demand for cattle products, mainly caused by increased affluence in developed countries and increased population in developing countries. Some of these technologies, such as milking cows by machine, had been invented many years ago (in about 1860 for the milking machine) but had remained unused until there was a ready market for the new technology. The demand for improved techniques of cattle breeding led to some pioneering discoveries that paved the way for the development of artificial methods of controlling reproduction in humans, once the ideas had been accepted in livestock. The first major development was artificial insemination of cows with stored bulls' semen. Even with the development and commercialization of embryo transfer techniques, it is artificial insemination that has undoubtedly had the greatest impact on breed improvement, and this has been much greater in cattle than in other livestock sectors.

Some of the major cattle breeds of the world

Aberdeen Angus

The Aberdeen Angus is a small, early-maturing breed of black cattle (Fig. 5.1) which was developed from naturally polled and small horned cattle that existed in Scotland in prehistoric times. It is used exclusively for beef production and produces a marbled flesh with good eating qualities. The breed was improved in the 18th and 19th centuries, when naturally polled cattle were selected because they are easier to handle. The absence of horns is a dominant trait, so ensuring that the offspring of first crosses are polled. Its small size and early maturity makes it particularly suitable for fattening autumn-born calves off pasture at 18 months of age.

Fig. 5.1. Aberdeen Angus cow (courtesy of Aberdeen Angus Cattle Society).

Ayrshire

The Ayrshire is a specialist dairy breed of cattle that is usually brown and white, which is kept mainly in Scotland, Scandinavia and North America (Fig. 5.2). It was developed initially in the 18th century by crossing black Scottish cattle with short-horned cattle of Dutch origin, West Highland cattle and Shorthorns. It is slightly smaller than its main rival, the Holstein–Friesian and produces less milk, hence its declining popularity. The fat and protein content of the milk is, however, slightly greater than that of the Friesian.

Belgian Blue

The Belgian Blue breed has its origins in the Belgian red or black and white pied cattle, which were crossed initially with Friesians and Shorthorns in the second half of the 19th century and later with Charolais cattle at the beginning of the 20th century, when the breed was officially formulated. The cattle are large, females usually weighing about 750 kg and males in excess of 1200 kg, and are used for intensive beef production. The breed is unique for its high proportion of cattle with prominent (double) muscling in the hindquarters. This characteristic has been genetically identified on the *mh* locus of a chromosome, theoretically enabling it to be transferred to other breeds. The upsurge in demand for lean beef in the latter part of the 20th century has created a rapid increase in demand for these cattle, which are now quite widely used to improve the beef production of the offspring from the extreme dairy type of Holstein cow and also in hill suckler herds.

Fig. 5.2.   Ayrshire cow (courtesy of the Ayrshire Cattle Society).

Brown Swiss

The Brown Swiss is one of the most ancient breeds of cattle still in regular use, with evidence of small red cattle being in Switzerland as early as 1800 BC. A medium-sized cattle breed, they now have a grey–brown coat, and have remained a dual-purpose milk and meat breed, with special application in high altitudes. Their short hair is complemented by pigmented skin, which protects against radiation when they graze at high latitudes or in the tropics. Their long period of development has produced a breed that is better than most dual-purpose animals at both milk and meat production, and the breed is used in many countries, especially Central and Eastern Europe.

Channel Island breeds

This is a collection of fine-boned, small cattle breeds that have been developed intensively for milk production. The milk from Channel Island cows contains more fat and, to a lesser extent, protein than other dairy breeds. The most widespread breed is the Jersey, whose females weigh only about 400 kg. They are normally fawn-coloured, although they can range from dark brown to almost white, with the extremities becoming gradually darker at the tips. The Jersey was developed at least in part from the breeds of north-west France, but since 1789, a reservoir of pure stock has been maintained on the island of Jersey by forbidding cattle imports. This has enabled it to resist diseases such as tuberculosis. Its use has been widespread in tropical regions because of its resistance to heat stress, but Jersey cows have the disadvantage of being particularly susceptible to hypocalcaemia. The cattle mature quickly, are docile and easy to handle, and can consume large amounts of forage for their size. They are unrivalled as producers of milk fat, but nowadays this is less valued than milk protein, as much of the milk sold for liquid consumption has some of the fat removed to increase consumer appeal.

Charolais

The Charolais is the most important French breed for beef production. It attained worldwide popularity in the 20th century, because of its large size and rapid growth rate. The Charolais cattle were developed from cattle imported into France during Roman times, and they are heavy boned since they were primarily used for draught purposes, with meat production of less importance. Mature cows weigh 800–900 kg and bulls in excess of 1200 kg. They are white or cream and the skin is light brown (Fig. 5.3), giving some resistance against sunburn. They are slow to mature compared with the British beef breeds, and are most suited to fattening at 24 months of age with some supplementary feed. The Charolais bull can be used to sire dairy cows, but calving difficulties are likely if they are used with small heifers. The crossbred calves are born up to 3 days later than Hereford calves and are about 4 kg heavier. Charolais cattle have less subcutaneous fat than the British beef breeds and are therefore well suited to modern requirements for lean meat. The potential to grow to a large size also suits today's market, as the initial investment in the calf can be utilized to the full in producing a large animal for slaughter and processing.

Fig. 5.3.   Charolais cow and calf (courtesy of the British Charolais Cattle Society).

Holstein–Friesian

The Friesian cattle came originally from the north-west of Holland, and in particular Friesland. This century, the cattle from this region have been developed into the highest-yielding dairy cow breed in the world. Nearly all cattle are black and white (Fig. 5.4), although a few are red and white. They are now in widespread use in most intensive dairying systems throughout the world, comprising for example 90% of the UK dairy herd. Both the Dutch and British Friesian cattle were until recently considered dual-purpose, but intense selection for milk production in America in the 20th century produced a strain called the American Holstein, which is probably a corruption of the word 'Holland'. The recent intermingling of Friesian and Holstein cattle in many countries, including the UK, has led to the breed often being classified as Holstein–Friesians. The American Holsteins are taller than Dutch Friesians and weigh 750–800 kg, compared with 650 kg for a traditional Dutch Friesian. New Zealand Friesians are even smaller and have a high capacity to produce milk from forage. In most Friesians and particularly Holsteins, the fat and, to a lesser extent, protein contents of their milk are low, although there have been recent efforts to increase these by breeding. Even though milk production efficiency is high because of the high yield potential, when the milk yield is corrected for solids content they are no more efficient than other extreme dairy breeds, such as the Channel Island breeds. However, their large size is beneficial in reducing the labour requirement per litre of milk produced, as labour input is largely a function of the number of cows on a farm. In countries with integrated milk and beef production systems such as the UK, the loss of meat production potential is significant.

Fig. 5.4. Holstein–Friesian cow (courtesy of Holstein–Friesian Society/Avoncroft Breeders Ltd).

Hereford

Although the Hereford was originally used both for milk production and draught purposes, as well as meat production, the Hereford is now exclusively used for meat production. It was developed in the county of this name in England, and early breed improvement in the 18th century produced an animal that matured early and would fatten off a pasture-based diet, but was still well suited to the demands of traction. Its docile character was useful for those working the animals in the yoke, and another attribute that was incorporated into the breed early on was the distinctive colour marking (red/brown body and white face, chest, bottom line, tail switch and feet; Fig. 5.5). The white features are dominant and enable farmers to recognize the breed of crossbred calves, giving them confidence when buying stock that they will have good fattening ability. In the 1950s and 1960s, the breed was developed into a smaller, more stocky shape that was suited to producing joints of beef of manageable size. The relatively early maturity of such cattle enables them to be finished off pasture at about 18 months of age. More recently, the breeding emphasis in the UK Hereford cattle has been for larger animals, with the increased tendency to finish cattle at a heavier weight.

The Americans developed a polled Hereford from the early 1900s, and they also increased its size, which improved growth rate. Importation of the polled American Herefords has led to 60% of registered Herefords in Britain now being polled. The inferiority of Hereford cattle to cattle of continental European breeds in slaughter weight was largely responsible for its reduced

Fig. 5.5.   Champion Hereford bull, cow and calf (courtesy of the Hereford Cattle Society).

popularity in its native England towards the end of the 20th century. With the emphasis now turning to less intensive feeding and ease of management, the Hereford is regaining popularity at home and is still a popular breed worldwide.

Limousin
The Limousin breed was developed in relatively harsh conditions in central France. The cattle are an orange–brown colour, with short legs and a large, well-fleshed rump. The cattle are, however, of nervous disposition and consequently can be difficult to handle. They are smaller than the other major continental breeds (the Charolais and Simmental), with cows normally weighing about 600 kg. Originally developed for dual purposes – draught and meat production – beginning in the 16th century, it has been extensively improved for meat production purposes over the last 150 years. It has recently been exported to many European countries, as its low level of subcutaneous fat and high potential for growth suits modern requirements. Towards the end of the 20th century, it became economically and technically feasible to feed high-quality supplements and high-energy maize silage to cattle, making it easier to finish late-maturing cattle such as the Limousin fast enough to be profitable. The Limousin is particularly suited as a crossing sire on Friesian cows, as the calves are relatively small at birth and there are very few calving difficulties.

Longhorn
It was the Spanish Longhorn that was exported to America, where it thrived on poor quality pasture as a slow-growing meat-producing breed. Its ability to calve without assistance was, and still is, valuable in the extensive ranches of America, but continental European breeds are now becoming more popular. The Texan Longhorn was initially improved by crossing with the Shorthorn,

and more recently the Hereford and Aberdeen Angus. The ability of the Longhorn to deposit most of its fat subcutaneously may again assume importance, as the fat can then be rapidly separated from the meat. Its ability to calve without assistance is valuable in extensive systems, but its low growth rate compared with cattle of continental European breeds will deter all those seeking profitable beef production from medium- or high-intensity systems.

Simmental

The Simmental is probably now the most popular dual-purpose cattle breed for milk and meat production. It originated in the Simme Valley in Switzerland and is now spread throughout Central and Eastern Europe. Originally they were also used for draught purposes, hence they are large, sturdy cattle with heavy bones (Fig. 5.6). This helps them to graze mountain pastures, where fine-boned cattle have reduced life expectancy because of their inability to cope with the harsh conditions. The Simmental is recognizable as various strains, such as the Swiss Simmental, Austrian Simmental and Fleckvieh (German and Austrian Simmentals). All are red or red–yellow and white, with the head being predominantly white. Although good milking cattle, Simmental-cross cows have been popular in recent times for suckling purposes, producing high-quality calves for meat production. As a crossing sire for dairy cattle, the Simmental produces calves that are slow to mature, indeed the bullocks usually require 24 months to finish even with supplementary concentrates. In this respect, the Simmental is similar to the Charolais, which is only slightly larger than itself.

Welsh Black

Although not widely known outside Europe, this breed is a good example of a modern suckler cow, as its small size is beneficial because the maintenance costs of the dam are kept low (Fig. 5.7). Traditionally dual-purpose, the Welsh Black has good milk producing characteristics which are useful for suckler

Fig. 5.6. Simmental steer cattle (courtesy of the British Simmental Cattle Society).

Fig. 5.7. Welsh Black cow.

cows, and the breed is particularly hardy, enabling it to thrive on upland pastures of poor quality.

Zebu (B. indicus)

Zebu cattle evolved from *B. namadicus* cattle in India, but have since been taken by man to South America, Australia and Africa, where their heat tolerance and natural resistance to tropical diseases enables them to thrive in areas where *B. taurus* cattle suffer high disease susceptibility. They generate less heat than *B. taurus*, partly as a result of their low productivity, and are characterized by a single hump on their back (Fig. 5.8), which allows fat to be stored in a concentrated reservoir, rather than subcutaneously over the whole body. This facilitates heat loss, as does their large surface area relative to their body volume. The large surface area is achieved by having folds of skin in the dewlap and preputial sheath, large ears and long thin legs. These effective cooling mechanisms allow the cows to maintain production at extreme temperatures and the bulls to remain fertile when the proportion of viable sperm in *B. taurus* cattle is severely diminished because of the heat. The hair of zebu cattle is short, sleek and often white, allowing the sun's rays to be reflected, and the underlying skin is usually pigmented to prevent cancers, particularly around the eyes. Their behaviour is unpredictable and they have a lively temperament, making them difficult to handle, but they survive well in extensive grazing conditions where little handling is necessary. Their breeding performance is not as good in their native environment as that of *B. taurus* cows in temperate conditions. They take longer to reach puberty, have long gestation lengths and postpartum anoestrus (Chenoweth, 1994; Dobson and Kamonpatana, 1986). When oestrus does occur, it is short and less overt. However, the cows tend to live longer than *B. taurus* cows and have strong

Fig. 5.8. Zebu bull, showing the hump, extended dewlap and preputial sheath.

maternal traits, hence their reluctance to release their milk without a calf being present. Normally the calf is tied at the head of the cow, while milk is taken by machine or by hand.

Two of the most common types of zebu cattle are the Brahman in India, and more recently the USA and Australia, and the White Fulani of the Sahelian region of Africa. Crosses with *B. taurus* breeds, which have produced such breeds as the Brangus, Braford, Santa Gertrudis and Droughtmaster are popular in intermediate climatic conditions.

Breed improvement

Cattle farmers that take a particular pride in the quality of their stock are often members of a breed society. The society records the details of all animals born on each farm and usually publishes records of all cattle in the society annually. These details include the date of birth of calves, their parentage and society number. Such cattle are called pedigree because their ancestry is known. Herd registers may be 'open', in which case non-pedigree cattle can be admitted, provided they meet the breed requirements for the colour and type of animal. Herd owners may be required to 'grade-up' their cattle by using a pedigree bull over a prescribed number of years. Alternatively, if a herd book is closed, it will only admit cattle to its register if both parents are also registered pedigree members.

Pedigree cattle attract a premium compared with non-pedigree or 'commercial' cattle. The premium is largely dependent on the breeders being able to demonstrate high performance in their stock to potential purchasers and they, therefore, must look after the animals well. The cattle are usually fed a high-quality diet to maximize performance, either milk production or growth rate, and this is not necessarily the most economic diet if milk or meat, rather than breeding stock, were the only outputs from the farm. High-quality diets allow the cattle to express their genetic potential but this often does not represent commercial practice. However, under all but the most extreme conditions, the ranking of the cattle for performance criteria will not be affected by the quality of diet, and pedigree cattle that perform better than commercial cattle on a high-quality diet will do the same on a low-quality diet.

Breeding objectives

Beef cattle

The major cost of beef cattle production in most systems is the food, which comprises about 80% of the variable costs. Therefore, improving the efficiency of beef cattle production should be directed at reducing the feed conversion ratio[1]. Beef cattle improvement has always suffered from an inability to measure one of the most important parameters that affects the efficiency of feed conversion,[1] that is the feed intake (Archer *et al.*, 1999; Korver, 1988). Breeders select on the basis of weight gain, usually to a fixed age of 400 days. This favours animals with high mature weights, since they have not started to divert nutrients to fat stores, which require more energy per kilogram live weight gain. The apparent increase in efficiency of the late-maturing breeds from continental Europe is an artefact of the 400-day test, but is only partly responsible for their increased popularity. Although these continental breeds, such as Charolais, Limousin and Simmental, are not inherently more efficient in their conversion of food to meat, they do allow farmers to take their cattle to higher weights and thereby dilute the high price paid for a calf per unit weight. The increase in mature weight tends to increase birth weight and may lead to increased calving difficulties when such animals are crossed with small dairy cows, such as first-calving heifers.

Within a breed, animals with a high ratio of muscle to other tissues at a certain stage of their development are likely to have increased value. In the last 25 years, a genetic mutation has appeared in the Belgian Blue breed, which increases the rate of muscle growth. The recent increase in popularity of the Belgian Blue cattle suggests that this 'double-muscle' trait will become part of future beef breeding programmes (Arthur, 1995). Double-muscled cattle have a considerably increased ratio of muscle to fat, particularly in the male, and they have smaller organ weights. The increase in size of the muscles is accompanied by an increase in tenderness, which increases the commercial value of both fore and hindquarter cuts. The trait is controlled by a major gene,

---

1   Food conversion ratio is food dry matter (DM) intake/live weight gain. Food conversion efficiency is the reciprocal of this, i.e. live weight gain/food DM intake.

which was recognized in some sheep breeds long before its recent appearance in the cattle. However, in Belgian Blue cattle, the trait is associated with difficult calvings in pure-bred animals, and such animals require specialist management to keep the number of Caesarean births to a minimum. The increased cost of care during calving may be financially justified by increased growth rate potential, but the practice arouses concerns for the welfare of the cows and may be the subject of future legislation or other control.

Apart from growth characteristics, future beef breeders are likely to concentrate on fat distribution and composition. Intramuscular fat is likely to be minimized because of health concerns, even though it imparts good cooking, and to some extent, eating characteristics. Subcutaneous fat will be more acceptable, although not in hot climates because of its effect of reducing an animal's ability to lose heat, and it is likely that protected supplements will be fed to some cattle to increase the deposition of mono- and polyunsaturated fatty acids that are beneficial for human health. It is likely that large breeds will be favoured, so that they can be slaughtered early while they are laying down mainly muscle tissue, not fat, but will have grown to be a reasonable size. Intensive, rapid finishing systems are likely to be rare as a result of demand for high-quality feeds for animals or humans that can use them more efficiently. Cattle that can survive on extensive grazing with little management will be favoured in industrialized countries, but in developing countries, it is likely that there will be intensification of grassland use to provide cattle for export and some home consumption. The low reproductive rate of many beef cows will encourage breeders to include this characteristic in breeding programmes, even though it partly derives from undernutrition (Opsomer and deKruif, 1999) and suckling (Veerkamp, 1990).

Dairy cattle

Selection for high milk yield has been the major emphasis of breeders in the past, which has tended to favour large dairy cows, such as the Holstein–Friesian breed. The physiological limits to production do not appear to have been attained, as cattle do not produce much more milk per kilogram body weight than other mammals. The cow does, however, lactate for a greater proportion of its life than most other mammals, so the stress of prolonged lactation may be considerable. The short lifespan of most cows in intensive dairy systems (normally about 5–6 years) is testament to the stress that they have to endure, when one considers that cows in less extensive systems often live to 20–25 years of age before they show signs of senescence. One of the deleterious effects of breeding for high milk yield has been to exacerbate the peak lactation, and it is the short period from the commencement of lactation to peak lactation that contains a high risk of metabolic breakdown, largely because intake is insufficient to provide the major nutritive requirements of the cow, and body tissue is mobilized at a high rate. Dairy cow breeders will probably focus in future on extending the lactation of cows, to diminish early lactation metabolic problems. It will only be possible to use cows with extended calving intervals if it is not desired to concentrate calving into certain periods of the year.

The inverse genetic correlation between milk yield and the concentration of many solids components of the milk, particularly fat, resulted in a decline in milk solids content as milk yield increased in the third quarter of the 20th century. This was undesirable as the transport costs of milk solids increase as more water is contained in the milk. In the last two decades, this has been addressed for some milk constituents by breeders, so that milk fat content of the Friesian cow, in particular, has increased from below 3.5% to nearly 4%. Recently, milk fat has become less valuable than milk protein in the industrialized countries, and maintaining a high milk fat content is now a low priority.

Reducing milk lactose content could be a beneficial objective in countries where lactose intolerance is common. However, the production of the enzyme lactase, which is required for lactose digestion, depends on the amount of milk that is consumed. People that traditionally consume large quantities of milk, such as those in Scandinavia, have little problem with lactose intolerance in adulthood, because they retain lactase activity. However, although lactose malabsorbers represent 90% of the human race, with nearly 100% of people from the Far East, 73% of USA Negroes, 42% of French, 20–30% of Britons and 3% of Scandinavians, most people can consume small quantities of milk, if they are introduced gradually into the diet, without any difficulty in digesting the lactose.

For infant feeding, it would be useful for cows' milk to be more similar in composition to human milk, with the ratio of cow to human milk solids contents currently being 3:1 for protein, 7:1 for casein, 1:1.6 for lactose. Herds may, in future, specialize in producing milk for certain functions. For example, cows' milk contains the protein lactoglobulin, which is hard to digest in liquid milk and is not present in human milk. The modern techniques of gene identification and transfer may enable the regulatory genes to be identified and switched off. Other possible specialized production systems could include increasing the casein content, to increase its value for cheese production, and increasing the phosphate content, which would increase the stability of the micelles and also calcium uptake. The ability to modify the mineral content of milk is reduced for some elements by the mammary glands homeostatic mechanisms, which closely control the concentration of the key mineral nutrients in milk, such as calcium, iron and sodium.

The tools for dairy cow improvement are well developed. Modern selection methods rank cows on profitability indices, which compare the margin over feed and quota costs for each daughter in each lactation. A typical formula for the profit index would give the highest weighting to protein production, less to fat production and only minor value to milk yield. These weightings are determined as the predicted transmitting ability (PTA), which is the predicted level of production that an animal is capable of passing on to its offspring above or below a predetermined baseline. In the UK, the baseline is updated every 5 years to reflect national herd improvement and the PTA value is accompanied by the year of calculation e.g. PTA 95.

More advanced profit indices are available, which include type characteristics with important economic merit (Veerkamp, 1998) and association to longevity (Essl, 1998), such as the profitable lifetime index, in which the milk production value of a bull's offspring is weighted for a lifespan value. The milk production value is the margin over feed and quota costs per lactation, determined annually. The lifespan value is calculated from conformation traits, such as good feet and udder, which are given a PTA. Thus, indices used by farmers increasingly compare cows for lifetime performance, because the value for individual lactations is not representative of the lifetime yield, which has more economic significance. In this way, disease susceptibility is incorporated into breeding programmes. However, the genetic correlation between lifetime performance and milk yield in a particular lactation is often less than 0.3. Management traits may assume increased importance on individual farms with specific problems, with calving ease, milking speed and temperament being three of the most important, because of public concerns for the welfare of dairy cows. Disease traits may, therefore, have a more important place in the setting of breeding objectives in future (Emanuelson, 1988). Under US conditions in 1994, the economic weighting, relative to a milk yield value of 1, of disease traits was as follows: clinical mastitis 0.25, laminitis 0.11, interval from calving to conception 0.10, milk fever 0.09, ketosis 0.07 and displaced abomasum 0.04 (Strandberg, 1996). However, this ignores the significance of diseases in contributing to poor welfare. An antagonistic relationship exists between at least one disease incidence trait, mastitis, and milk yield, making it difficult to achieve progress in both traits. Unfortunately, although disease incidence traits are economically and ethically important to be included in breeding objectives, the heritabilities of the traits are often low.

Reproductive traits usually have a heritability of less than 0.10 and again there is an antagonistic genetic relationship with milk yield. Longevity, or stayability, as it is more correctly known, also has a low heritability, about 0.05 (Strandberg, 1996). The ability to include reproductive and disease traits in a breeding index depends on what measures are taken. Somatic cell counts (SCC) can be used to indicate mastitis severity, but caution should be exercised in breeding cows with low SCC that may not respond well to bacterial challenge. Calving to first service interval could be included as a reproductive trait, but it must be remembered that the more traits included the less progress will be made in any individual trait. If no health or reproduction traits are measured, longevity in the herd could be included, but it should be adjusted for production.

Type traits may be included in a breeding index, but in most cases little is known about their correlation with economic functions, so they are most likely to be used by farmers to correct traits that they feel are particularly deficient in their herd. Previous selection may have had adverse effects on type traits, for example, selection for milk yield alone results in a deterioration of udder characteristics. Body traits tend to have heritabilities of about 0.25–0.45, udder traits 0.2–0.3 and feet and leg traits 0.15–0.25. It is difficult to relate type traits recorded in different countries, but a commonly used recording system is not yet available.

The newest dairy cow improvement schemes are attempting to locate regions on the 30 pairs of bovine chromosomes that are responsible for a significant impact on profitability – known as quantitative trait loci (QTL). Other factors that could be selected for include welfare parameters, such as disease resistance, or reproductive success, which could enable the recent decline in dairy cow fertility to be overcome. The detection of QTLs will enable specific traits to be selected for much more rapidly in the progeny of a particular bull. Blood or hair samples can be used to obtain the genetic material required for detecting QTLs. Multiplying up the selected genes is theoretically achievable by cloning, but care has to be taken that this does not erode the genetic diversity of the parent stock. At present the origin of the genetic donors still largely determines whether they successfully develop or not, with those from the embryo being most likely to succeed. Since high-producing cows are generally more profitable in a range of environments, it is likely that further concentration on high-producing Holstein cows will follow a widespread use of cloning techniques. It is true that in certain environments, particularly those that are hostile to high milk output per cow, locally evolved genes can increase the ability of carriers to thrive in the environment. This has been demonstrated in tropical climates, where the disease resistance provided by locally adapted breeds enables them to perform better than imported cattle that were bred for high output temperate conditions. Often the benefits of both types are realized by crossing imported cattle with local breeds, with the greatest benefits of hybrid vigour being afforded to the first generation (F1) crosses. Attempts to maintain a permanent stock of crossbred cattle have met with some success, spawning new breeds such as the Brangus (Brahman cross Aberdeen Angus).

Calving Patterns

A cow's breeding cycle is usually controlled by the farmer, either by limiting access to the bull or by the cow being inseminated artificially with semen collected from a bull, stored in liquid nitrogen and injected into the cow's reproductive tract with a syringe. Thus, providing the farmer knows when a cow is at the right stage for insemination, he can control when she breeds. In some countries, such as New Zealand, where most of the milk is used for manufacturing purposes, dairy farmers use a strict block-calving pattern. Most of the cows will calve in a period of 2–3 months in spring, so that more milk can be produced from grass in summer and the amount of supplementary feed required in winter is kept to a minimum. It also gives the farmer a quiet period at Christmas time when most of the cows are dry (non-lactating).

In Britain, block-calving patterns are also popular because they allow synchronization of the major activities, but the need to produce milk all year round for the liquid market and to keep the factories for milk products working at a uniform rate all year has led to calving being focused on three times of the year – autumn, spring and summer (Fig 5.9). Milk price incentives for producing milk in late summer have been introduced, since pasture is usually of poor quality and quantity at this time. Farmers that block calve their cows in early summer usually have to feed supplementary forage and concentrates when they are in mid-lactation in late summer, and in some systems cows calving in mid-summer remain inside after calving to be fed on conserved feeds. Farmers have to judge whether the increased cost of production is justified by the increased milk price. Changing the calving pattern may take several years, so the decision needs to be made with long-term price forecasts in mind.

Block calving has several advantages:

1. Management is simplified, since efforts can be targeted to certain activities at certain times of year, e.g. calving, oestrus detection. In relation to oestrus detection, it is advantageous to have several cows likely to come into oestrus at one time, as the chances of them forming into a sexually active group are increased.
2. There is a quiet time when all the herd is dry. This may be more important for small herds run by one person, where there are no additional staff that can be left in charge.
3. There is more incentive to keep to a tight calving schedule, as cows which calve outside the main calving period may have to be culled.

However, there are also disadvantages of block calving:

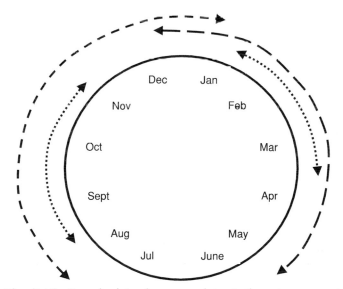

Fig. 5.9. The distribution of calving for cows calving in the autumn or spring in either a blocked (<·········>) or spread (← – – – →) pattern.

1. There are peaks of labour requirement, e.g. at calving time, which may stretch the farm's resources.

2. Cows may have to be culled unnecessarily if they calve outside the main calving period, particularly older cows with high milk yields but longer than average calving intervals. As the mean calving interval is about 395 days in Britain, an average cow that calves for the first time at the beginning of a 3-month period can only survive for four lactations before she calves outside the main calving period of the rest of the herd.

3. Heifer rearing is constrained into either a 2- or 3-year period to first calving, when some farms might like to calve their cows at about 30 months of age. The first calving must be planned for the very beginning of the calving period, particularly if the farm's average calving interval is long.

4. A 365-day calving interval, which would enable cows to remain in a tight block-calving pattern indefinitely, may not be optimum for high-yielding cows. These cows may have to cease lactating when giving over 10 l day^{-1}. For low-yielding cows, a 355-day interval gives the maximum milk yield (Fig. 5.10).

Given the difficulty in obtaining high milk yields from grazing cows in late summer, it is particularly important for farmers with spring-calving herds to operate a tight block-calving system. Unless large amounts of supplementary feed are fed in August and September, cows that calve in April and May will have low lactation yields (Fig. 5.11). Farmers with summer calving cows may have a deliberate policy of starting to feed conserved forages early, but this will often not be the case for the remnants of a spring-calving herd. In the British Isles, spring-calving herds predominate in the west of the country, where rainfall is higher and farmers want to make the best use of grazed grass possible. Herds in this part of the country have less reliance on purchased concentrates, but fertility management (and grazing management) needs to be good. Conception rates are naturally reduced in late spring/early summer because the cows would then calve in mid-winter, with peak feed requirements soon after, which would not be a good evolutionary strategy to ensure

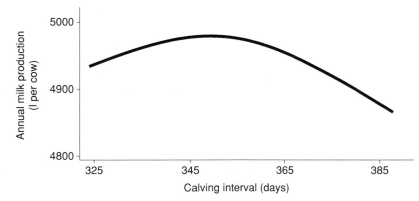

Fig. 5.10. The relationship between the annual milk production and calving interval for cows giving a mean milk yield of almost 5000 l during the lactation.

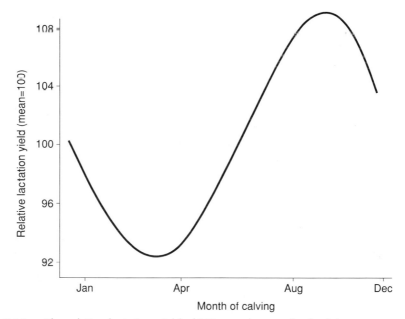

Fig. 5.11. The relative lactation yield of UK cows by month of calving.

calf survival. In summer, it is more difficult to get cows pregnant during hot weather (more than about 30°C, depending on the humidity), because oestrus is short and of reduced intensity and it is more likely to be expressed at night when it is cooler. Conception rates are also reduced. In winter the ambient temperature has to fall below −10°C before reproduction is adversely affected, which is mainly because mounting activity is surpressed.

Artificial Insemination

Artificial insemination may be preferred to insemination by a bull, because the rate of genetic progress can be increased, there is no cost or danger associated with keeping a bull on the farm and the conception rate may be increased. To achieve genetic progress in a herd, a bull must be proven to be of high potential by test-mating it with at least 20 cows. The farmer then has to wait 4–5 years until the performance of the offspring is known. Even if the bull is proven to have high genetic merit, he can only be used for a maximum of four matings per week, giving him limited reproductive capacity compared with 30,000 matings per year that are possible when a bull is used for artificial insemination.

The main disadvantages of inseminating artificially are that a skilled operator is needed at the precise time of the cow's cycle when conception is highest. This may not cause a difficulty in most developed countries, but in many developing countries, it is difficult to organize because of poor road links to farms, irregular communication channels and a lack of skilled labour. Providing the necessary resources for artificial insemination is relatively more costly in developing countries, especially if the transport and semen storage costs are high. A second disadvantage of artificial insemination is the potential loss of genetic diversity caused by farmers using a small number of high-value bulls. This could create risks in the future if priorities for breed development change. For example, the priorities at present focus in most countries on increasing the genetic potential for output of milk solids per cow, in particular milk protein. This can only be achieved if additional high-concentrate food is provided to meet the extra nutrient demands. However, such food may in the future need to be reserved for human consumption. Flexibility in the future breeding programme is lost irreversibly by restricting genetic resources to animals capable of fulfilling a limited range of objectives. A further disadvantage of artificial insemination is that it is necessary to know precisely the stage of the oestrus cycle of the cow in order to inseminate her at the right time. Previously this has usually been assessed by the herdsperson watching their cows regularly to determine when they stand to be mounted – a definite sign that they are in oestrus. The increased number of cows that most herdspeople are now required to attend to in profitable units gives them insufficient time to stand watching for signs of oestrus. This may be why several scientists have reported reductions in dairy cow fertility in recent years. It is much more difficult for the herdsperson to detect oestrus than for a bull, who can detect that a cow is preparing to enter oestrus up to 4 days before the event.

Accurate timing of the artificial insemination requires the herdsperson to know when a cow enters oestrus, since the maximum chance of conception occurs at 12–24 h after the onset of oestrus. Oestrus lasts normally for 12–16 h and ovulation occurs approximately 12 h after the end of oestrus. Many farmers use the a.m./p.m. rule to determine when to inseminate their cows, providing that they have the choice of inseminating cows after the morning or afternoon milking: cows that are seen in oestrus during the morning are inseminated in the afternoon and those that are still in heat in the afternoon are re-inseminated the following morning. Those that are first seen in oestrus during the afternoon or evening are inseminated the next morning. If a farmer only has the possibility to inseminate once a day, cows seen in oestrus before the morning milking should be inseminated that day, otherwise they should be inseminated the following day.

Artificial insemination technique

Artificial insemination requires semen to be collected from a bull, who is encouraged to mount a dummy cow or sometimes a 'teaser' cow and his penis is manually directed into an artificial vagina. This has a heated jacket to maintain the device at the right temperature and to ensure that the conditions in the rubber sheath replicate the conditions of the vagina as closely as possible. Attached to the end of the rubber sheath is a small phial in which the semen is temporarily stored before being transferred to a more permanent storage facility after the addition of a diluent, which contains egg yolk, water, sodium and potassium salts and antibiotics. The diluent prevents temperature shock, buffers the pH and increases sample volume about 100 times. Permanent storage is achieved in the frozen state in liquid nitrogen in a flask that can be transported from farm to farm by the inseminators. When an inseminator visits a farm, he or she expects the recipient cow to be ready and restrained. Semen is then deposited in the uterine body using an insemination gun. The gun is usually protected from temperature shock by wrapping it in a towel and placing it inside the inseminator's overalls. One hand is then inserted into the rectum where it can manipulate the cervix, with the other hand used to guide the inseminating gun into the vagina. After guiding the gun through the cervical opening with the first hand and then into the cervical body, the gun is gradually withdrawn until it is flush with the cervical opening. The semen is then gradually released over a 5-s period to achieve good distribution. This positioning applies to all first services, but in subsequent services the cervical mucus may feel viscous and sticky, in which case the semen should be deposited halfway through the cervix in case the cow is pregnant, when penetrating the cervical plug of mucus would jeopardize the pregnancy.

The insemination can be done by 'do-it-yourself' operators on the farm, but because they are performing fewer inseminations than professional inseminators, the success rates are often lower. Do-it-yourself inseminators have the advantage that they can inseminate cows twice a day, whereas professional inseminators will usually only visit farms once each day.

Embryo transfer

The transfer of fertilized ova (embryos) from a donor female to recipient females is usually coupled with superovulation of the donor female and is known as multiple ovulation and embryo transfer (MOET) (Callesen *et al.*, 1998; Greve and Madison, 1991). The technique is principally used to increase calf production from the best cows, in the same way that artificial insemination is used to extend the number of offspring from high-value bulls. The technique can also be used to induce twinning, to speed up a selection programme or to export stock to developing countries without the attendant problems of adaptation of cattle to hostile environments after growing up in temperate conditions. In future, if the technique becomes more successful, it could be used to obtain high-value beef cattle from low-quality dairy cows. It could also be used in conjunction with cloning and embryo-sexing techniques once they are established procedures in commercial cattle breeding.

The technique has had variable results, with the average number of transferable embryos from a superovulated dam being about five, but with considerable variation around this mean. This is largely caused by variation in the ovulation rate, but also the variations in fertilization rates of the ova and in the recovery rates are significant (Gordon, 1994).

Management of the donor cow

The donor cow is superovulated with gonadotrophins, usually pregnant mare serum gonadotrophin (PMSG). This is long lasting and a single injection is adequate, but may need to be counteracted with anti-PMSG postovulation to prevent a second wave of follicles. Follicle-stimulating hormone (FSH) is an alternative but it has a much shorter half-life, requiring several injections over a period of 3 days. After superovulating with PMSG or FSH, a prostaglandin injection is given as a luteolytic agent.

Good results can be obtained when superovulating if the following conditions are adhered to:

- adequate nutrition and absence of stress;
- donors prepared so that they are superovulated in mid-cycle, possibly using an induced oestrus cycle;
- donors inseminated with high-quality semen at 12–24 h after the onset of oestrus behaviour. Bulls differ in fertility and best practice must be followed when inseminating. The insemination can be repeated after 12 h if the cow is still in oestrus.

Embryos are recovered about 7 days postovulation by flushing a sterile solution through the uterus. This is either introduced surgically, or in the case of milking cows, non-surgically via the cervix. With docile cows, a flank incision is possible, but for range cattle a ventral incision is most likely to be required together with a general anaesthetic. Any uterine damage resulting in haemorrhaging is likely to be embryotoxic.

Freezing the embryos

The first bovine embryos were successfully frozen in the early 1970s. Nowadays over one-third of recovered embryos are frozen and thawed before implantation. After recovery of the embryos, they are washed several times, including in trypsin to inactivate any viruses, and then frozen in phosphate-buffered solution, with added glycerol as a cryoprotectant. Freezing should start within 4 h of recovery. The embryo dehydrates as the temperature is lowered and this determines the optimum freezing rate – too fast and deadly ice crystals form intracellularly, too slow and the embryos become excessively dehydrated. Only top-grade embryos should be frozen. An alternative to freezing is to store the embryos in acid buffer solution for 24–36 h before implanting in the recipients.

Management of the recipient cows

The cycle of the recipient cows must be synchronized with that of the donor, at least within 24 h; this is often achieved by prostaglandin injections for both donor and recipient. If the embryos are frozen after being collected from the donor, the preparation of the correct number of recipients is facilitated. As with the donor cows, the recipients should be well nourished, in good health and free from stress. The embryos can be inserted either surgically or non-surgically. In surgical transfer of embryos that are at the morula to blastocyst stage, a small abdominal incision to expose the uterus is made, and the embryos transferred to the lumen of the uterus via a pipette or catheter tip. Younger embryos are inserted into the oviduct, allowing them time to develop before being exposed to uterine conditions. Non-surgical transfer is similar to artificial insemination, with the embryos being deposited one-third of the way up the uterine horn. Sterile procedures are essential with both surgical and non-surgical transfer, with particular care necessary to prevent faecal contamination of the reproductive tract in the case of non-surgical transfer, to eliminate any risk of infection.

Oestrus in the Cow and its Detection by the Herdsperson

Oestrus is the behavioural manifestation of sexual receptivity in the cow (the term derives from the similarity between this behaviour and that when they are attacked by the gadfly, spp. *Oestrus*). The behaviour lasts for on average 14 h immediately prior to ovulation in each 21-day cycle, and is longer in cows than heifers. It is enhanced by the presence of the bull, through a process known as biostimulation. The first oestrus signs are normally displayed about 40 days postpartum, although it can be considerably longer in some animals. Some cows undergo a 'silent heat' at the time of the first oestrus postpartum, in which ovulation occurs but there are no signs of oestrus.

Recognition of oestrus by the herdsperson is essential if the cows are to be served by artificial insemination. The definitive sign of oestrus is that a cow allows herself to be mounted by another cow. Cows will not voluntarily perform this activity outside oestrus, but they may be mounted against their will if they are unable to escape. Young heifers are particularly at risk of being mounted forcibly. Cows normally stand to be mounted about 50–60 times at each oestrus event, but for about one-quarter of cows, it is less than 30 times. It is therefore necessary for the herdsperson to be able to detect all the signs of oestrus and know at what stage of the oestrus cycle they are performed.

A typical oestrus cycle

The oestrus cycle can be divided up into the following periods:

- pro-oestrus, 10 h;
- oestrus, 14 h;
- metoestrus, 10 h;
- dioestrus, 19–20 days.

In pro-oestrus, a cow becomes restless and will separate herself from her associates if it is possible. She becomes aggressive to other cows to assert her dominance and right of access to a bull, if one is available. At the start of oestrus, she gathers with other cows at similar stage of the cycle to form a sexually active group. The larger the sexually active group the more exaggerated the sexual behaviour between the cows. Overt sexual behaviour may be shorter in a large sexually active group than a small one, because of cows becoming exhausted and satiated. The activities performed by cows in the group include interactive mounting, resting their chins on each other's backs and rubbing them up and down, sniffing and licking the genital region of other cows to detect the pheromones produced during oestrus, vocalizing and sampling the urine of other cows. At the start of oestrus, cows mount other cows but are not mounted themselves. In this way, grazing wild cows seem to have developed the behaviour over the course of evolution as a signal to bulls that would be grazing some distance from the matriarchal group. The behaviour indicates that there are receptive cows in the group. The mounting cow is not necessarily in oestrus but would expect to receive the benefit of the bull's attention once he had copulated with the mounted cow in oestrus. The bull's copulation capabilities are legendary and were responsible for cattle being deified in many ancient societies.

In the middle of oestrus, cows therefore stand motionless waiting to be mounted and will also mount other cows, often accompanied by pelvic thrusting that confirms the mounted cows' state of receptivity. A heightened electrical sensitivity of the vagina of the mounted cow at this stage of the cycle probably encourages her to perform the mounting behaviour. Vaginal mucus production increases and the herdsperson may see mucus emanating from the cow's vulval labia or on the back of their concrete cubicle bed. During this stage of oestrus, cows bellow repeatedly and are very restless.

In metoestrus, cows recover from their activities by lying and resting. They may eat more food, since eating can be temporarily suspended during oestrus. In dioestrus, there are no signs of sexual activity except occasional mounting, which may signify temporary build-up of the reproductive hormone oestrogen.

Oestrus detection

The oestrus detection rate for a herd of cows can be determined by dividing the number of cows observed in oestrus over a predefined period by the number of cows that would be expected to enter oestrus during this period. The latter is determined from a knowledge of which cows are cycling and assuming a 21-day oestrus cycle. The target detection rate should be 80%, but frequently it is less than 60%.

A significant problem in detecting oestrus is its variable length and intensity. About 20% of cows will have oestrus periods of less than 6 h, and in 5% of cows it is more than 18 h. Some of this variation is predictable, such as the reduced oestrus displayed by young cows. However, the herdsperson should always strive to provide an environment that allows cows to show their maximum oestrus display. A good floor for a firm foothold is essential, and a well-lit environment with few encumbrances is most beneficial. High stocking densities help to bring cows into contact with each other more often and stimulate sexual activity. Often oestrus is shown by grazing cows when they are collected together for milking. As suggested earlier, the size of the sexually active group is influential, with large groups promoting a short intense oestrus compared with small groups. However, the introduction of new members to the group may stimulate renewed interest in sexual activity in cows that have been members of the group for several hours. The size of the group will depend on the size of the group of cows from which the sexually active group can form and the proportion of the cows that are cycling at the time, which depends on the season and the spread of calving mainly. For example, in small groups of 10–15 cows it is unlikely that there will be more than two cows interacting sexually at any one time.

A typical frequency of the different behaviours associated with oestrus is shown in Table 5.1. Other signs that may be seen include the Flehman behaviour, mucus dripping from the vulva, and the cows' flanks being streaked with dirt and steaming after mounting activity. The Flehman behaviour is an up-curling and retraction of the upper lip and partial opening of the mouth to allow air to reach the vomeronasal organ, usually coupled with extension of the head in the direction of the odours. The vomeronasal organ is the cow's pheromone-sensing organ, accessed via two openings in the roof of the mouth, which is particularly used for exchanging sexual information. The herdsperson may also notice individual cows being more aggressive and bellowing. In tethered cows, the herdsperson is most likely to notice the restless nature of an oestrus cow, as well as mucus production from the vulva. The small number of tethered cows that each person can look after means that oestrus detection is often not as difficult as with large herds of loose-housed cows.

The short duration of oestrus in many cows makes it particularly hard to detect, unless the herdsperson is able to spend a considerable amount of time with the herd. The greater reliance on mechanical aids, increased need for detailed records and modern housing systems that mean that cows are now more likely to be managed as large groups rather than individuals, reduce the contact between the herdsperson and the cows within their care. The problem

Table 5.1. Typical frequency of oestrus behaviours (adapted from Phillips and Schofield, 1990).

Behaviour	Events per hour
Mounted with standing reflex[a]	3
Mounted without the standing reflex	1
Sniffing/licking the genital area	5
Chin rubbing/resting	6

[a]About one-half of these occur with pelvic thrusting.

is most acute soon after calving, when the herdsperson should watch for the first oestrus, so that he/she can be alert 21 days later for the oestrus at which the cow will probably be served. A weekly computer printout of cows that are expected to enter oestrus each day will assist in achieving a high detection rate. Cows should not be served before 42 days postpartum as conception rates are often less than 40%, whereas they increase to 50% by 50 days postpartum. There is, therefore, usually only one initial oestrus that should be recorded but no action taken. Many herdspeople believe that this first oestrus postpartum is usually silent. However, the true incidence of silent oestrus is only about 5%, but the reduced nature of the first oestrus leads to many displays being missed. Poor nutrition may reduce the oestrus display, but not sufficiently to make it difficult for the herdsperson to detect. Cows with high milk yields and old cows usually show a diminished oestrus display, and young cows that are afraid of interacting with, and being mounted by, older cows, also show less oestrus behaviour. Lame cows will obviously be less likely to mount each other. Oestrus is also less manifest for *B. indicus* than *B. taurus* cows, often leading to a reduced reproductive rate in the former. High temperatures depress oestrus display, and encourage cows to show oestrus at night when it is cooler. Some farms in developing countries employ people to watch the cows at night for oestrus and they are often paid on results. In Western Europe, it is traditional to have small herds, often run by families, but in recent years these herds have expanded, leaving the farmer with less time to look for oestrus. Under such conditions, oestrus detection rates are only likely to be acceptable (>80%) if the cows are watched three times a day for 15–20 min at a time. These occasions should include an observation before morning milking and one late in the evening. Cows do not show much oestrus behaviour when there are other activities taking place, such as feeding or milking. Ideally, oestrus detection should be a team activity, with everyone involved with the cows taking part. A blackboard hung in the milking parlour can be used by members of the team to note any cows that have been observed engaging in oestrus behaviour, and cases confirmed by the herdsperson should be entered into a computer record.

The importance of oestrus detection on dairy farms has led to a number of detection aids being developed. Most of these will only assist the herdsperson and cannot replace careful observation. The system with potential to avoid herdsperson observation altogether is the attachment of pedometers to the lower leg or around the necks of all cows that are expected to be cycling. These record the number of steps that a cow takes, which increases three- to fourfold during oestrus. The scale and predictability of the increase means that it can provide an accurate system of detection without herdsperson observation. The device records step number by a mercury switch, mechanical pendulum or piezo-electric member, and it usually relays the information to a data-capture unit in the parlour, although some transmit directly to a signal-emitting unit on the device itself. Most pedometers indicate that oestrus has occurred when the number of steps taken between two milkings doubles, but greater accuracy can be achieved with an algorithm based on information accumulated over several days. The opportunities for accurate insemination are greater when step number records are downloaded from the pedometers three times daily, but satisfactory performance can be obtained with twice-daily readings. Pedometers have the disadvantages that they are expensive to produce and can easily become detached from the cow's leg. Sometimes step numbers change for reasons not associated with oestrus – cows being lame or turned out to pasture, for example.

Other oestrus detection aids for the herdsperson include tailhead indicators, milk indicators, video cameras and cervical mucous conductivity measurements. Tailhead indicators are the most popular. These rely on the pressure that a mounting cow puts on the tailhead, which triggers either the release of a dye from a capsule stuck onto the tailhead, or the scuffing of paint put on the tailhead by the herdsperson.

Milk indicators include:

- milk progesterone concentration, which declines below 1 ng ml^{-1} for 3–5 days around oestrus;
- milk yield, which usually declines after the onset of oestrus because it is withheld in the mammary gland and then increases afterwards as the extra milk in the gland is released;
- milk temperature, which increases by about 0.1°C during oestrus.

The reduction in milk progesterone content is potentially useful if the point of decline can be determined, otherwise a single low value will not determine the stage of oestrus sufficiently accurately. There are kits available that indicate whether the progesterone content of milk is indicative of oestrus using a colorimetric test that takes about 20 min. The reduction in milk yield can provide associated evidence but the scale of the change depends on when a cow is milked relative to the time that she enters oestrus. Cows that are first milked several hours after the onset of oestrus have a significant decline in yield and milk fat concentration at the first milking, caused by milk being withheld in the mammary gland. There is usually a corresponding increase in yield at the next milking. The small increase in milk temperature at oestrus is difficult to record, particularly if it is measured after air has entered the system to cause temperature fluctuation. Subcutaneous implants may in future be able to overcome some of these problems, recording parameters in blood rather than milk.

Cervical mucus conductivity increases during oestrus, since more sodium ions are produced. This can be measured in the vagina, about 5 cm from the cervix preferably, using a probe with copper electrodes on the end. However, the variation caused by air pockets in the vagina and risk of introducing infection reduce the value of this method of oestrus detection.

Oestrus detection aids are useful but are no substitute for good stockmanship – which includes regular observation of cow behaviour, a knowledge of what to look for to determine the best time to serve cows and the provision of an environment that is conducive to sexual activity in the cows. The mounting display during oestrus is unique to cattle and may even have partly evolved in response to human selection. It is important that all stockpeople understand its significance and know how to create conditions that favour its exhibition.

Conception

Having served a cow at the correct time, it is essential that uterine conditions are optimal for conception and implantation to take place. The proximity of the service to calving is important, as indicated above. Feeding is the other major influence, with cows that are on a rising and high plane of nutrition being more likely to conceive. Low or negative weight gain is the major reason for low conception rates in early lactation cows. Each increase in weight gain of 0.1 kg decreases the calving to conception interval by 30 days on average. Cows should be in condition score 2–3 on the five-point scale, with those that are either fatter or thinner being likely to have conception rates less than 50%. Both energy and protein are important to consider, and some minerals, e.g. phosphorus, may play a part. The season of the year also affects conception rate, with cows served in early autumn having high conception rates as they would naturally give birth in early summer when there is plenty of grass available.

The conception rate can be estimated on the farm by calculating the proportion of cows served that are diagnosed pregnant 40–70 days later. For cows served naturally, the conception rate should be approximately 70%, for those served artificially about 65%, depending on the inseminator. Some bulls have low conception rates because of poor quality semen, so excessive use of one bull may be dangerous unless conception rates for other cows that have been inseminated with his semen are known.

Other important measures of reproductive rates, apart from the oestrus detection rate and conception rate, are the calving index (interval between calvings), the number of services per conception and the calving to conception interval. The calving interval is normally assumed to be optimal at 365 days. This minimizes loss of milk as a result of a prolonged dry period but still gives the cow time to recover if lactation is terminated at 305 days. Increasingly farmers are questioning the wisdom of stopping the lactation of high-yielding cows early. This is especially true if the value of an annual calf is less than formerly. Regarding the interval from calving to conception, if we assume that the gestation lasts for 280 days, then to achieve an annual calving at the same time each year the calving to conception interval should be 365 − 280 = 85 days. If the conception rate is 50% and the oestrus cycle 21 days long, then the cow must be first served at 65 days postpartum to conceive at, on average, 85 days postpartum. Working back still further, if a cow has to be served for the first time at 65 days postpartum, and she begins to cycle at 40 days postpartum, then there are only 25 days in which to observe oestrus i.e. just over one cycle. A herdsperson therefore can only afford to miss one oestrus before each cow needs to be served. This emphasizes the importance of seeing the first oestrus postpartum and waiting to see the next one 21 days later.

Hormonal Control of the Oestrus Cycle

The bovine oestrus cycle can be divided into the follicular and luteal phases. During the follicular phase, the ovarian follicle develops in response to the secretion of FSH by the anterior pituitary gland. The follicles produce oestrogen, which controls oestrus behaviour. After stimulation by oestrogen, the gland then produces a surge of luteinizing hormone, which triggers ovulation or release of the ovum from the follicle. The follicle turns into a corpus luteum, or yellow body, which produces progesterone to act on the uterus to produce a suitable environment for implantation and prevent further ovarian activity. If the ovum is not fertilized, the uterus produces prostaglandin which kills the corpus luteum, allowing progesterone levels to increase and FSH to act on a new follicle.

A herd where there are problems with cow reproduction should be first investigated for low oestrus detection rates. If they are low and oestrus detection aids fail to bring improvement in reproductive performance, and if no bull is available, the herdsperson may have no other option than to synchronize the cycles of non-pregnant cows with exogenous hormones, followed by artificial insemination. There are two main methods of doing this – either to administer prostaglandins that induce early regression of the corpus luteum, or to supply progestagens that act as an artificial corpus luteum.

Prostaglandins, originally thought to come from the prostate gland in men, will induce corpus luteum regression from day 6 to day 16 of the cycle. Two injections given 11–13 days apart will ensure that all animals will have a functional corpus luteum at the time of the second injection, providing that they were cycling originally. Following the second injection cows should come into oestrus 3–4 days later. The variation in time to oestrus is significant in adult cows, as a result of the stage of follicle development when the injections are given, and therefore they cannot be inseminated at a fixed time after the second injection. Dairy heifers can, however, as the variation is less, with inseminations at 72 and 96 h after the second injection giving good results. One insemination at 80 h is possible but insemination after observed oestrus is best. One disadvantage of using prostaglandins is that cows that are pregnant and are treated in error will abort. Similarly, pregnant women that administer the treatment are at risk if some is absorbed through the skin. Prostaglandins should be administered only after veterinary examination and by a veterinary surgeon, in view of the potency of the hormones in humans. An alternative use of prostaglandin is in the early postpartum period when it stimulates uterine involution and early return to oestrus cyclicity, but routine use on all cows at this stage in the reproductive cycle, rather than just problem cows, is not usually justified.

Progestagens mimic the luteal phase of the cycle and as such may be useful in both cycling and non-cycling cows. Treatment is required for a period of 10-12 days, usually administered by a progesterone-releasing intravaginal device and may be combined with pregnant mares' serum gonadotrophin to stimulate follicular development. Over-dosing with PMSG can lead to multiple births in beef cattle, which is usually considered undesirable. After removal of the silastic coil that is impregnated with the progesterone, an injection of prostaglandin is given 2 days later to ensure regression of the corpus luteum. This routine provides an accurate synchronization of oestrus and cows can be inseminated at a fixed time, 56 h after implant withdrawal (Odde, 1990).

Pregnancy Diagnosis

The importance for economic production of rapidly returning cows to pregnancy after they have given birth requires that the pregnancy is confirmed or otherwise as soon as possible after service. The earliest opportunity to test for pregnancy comes about 21 days after the cow was served, by investigating whether she exhibits another oestrus. Although this may be observed by the herdsperson as a behavioural oestrus, it can also be tested hormonally by measuring the progesterone content of the milk. If the cow is pregnant, the corpus luteum will produce progesterone, if she is not but she is in the follicular phase, the progesterone level will fall to less than 1 ng ml^{-1} over a period of 3–5 days around oestrus. A problem with this method is that the cow may have been served initially when she was not in oestrus, if for example she was incorrectly diagnosed by the herdsperson. Repeated samples could detect this anomaly, but are not always practical for large numbers of cows. A hormone produced by the cotyledons of the placenta, oestrone sulphate, can be used with some degree of success, but not until mid–late pregnancy. Some farmers diagnose pregnancy themselves using non-return to oestrus and the cow's body condition as indicators. They are only accurate about one-half of the time with non-pregnant cows. A small proportion of cows (about 7%) exhibit signs of oestrus during pregnancy.

An alternative to milk progesterone analysis is ultrasound examination of the uterus, which can be performed as early as day 26 after service and is quick and easy to do. The procedure, however, requires expensive equipment. The most popular method of pregnancy diagnosis is an internal examination of the ovaries via the rectum, otherwise known as rectal palpation. The initial signs of pregnancy – asymmetry of the uterine horns, the presence of fluid in the larger horn and the presence of the amniotic vesicle – can be detected at 30–35 days after service by experienced veterinarians. There is a risk of subsequent embryo loss if the palpation is too vigorous. At this stage and up to 65 days post-service the date of conception can be accurately determined. After 65 days, the amniotic vesicle becomes too flaccid to be recognized, but the fetus can be palpated and the date of conception determined to within a 1-week period using fetal head development as a criterion. In many cases, pregnancy diagnosis is conducted at 90 days post-service, by which time the uterus is flaccid and both the placentomes and fetus can be palpated.

Parturition

Parturition is a critical time for both cow and calf and is particularly difficult for a first-calving cow, because of her inexperience and the small size of the pelvis through which the calf has to pass. Other risks include *pre partum* damage to the fetus, failure of the neonatal calf to maintain its immune status and the stealing of the calf by other cows or, in rangeland conditions, predators. Many inexperienced herdspeople are eager to give as much help as possible, but patient observation is the best assistance in most cases. Normally a grazing cow will retire from the rest of the herd and find a quiet, sheltered place to give birth, such as under a hedge. The herdsperson should recognize impending calving and, if difficulties are anticipated, it is better to bring the cow to a calving box before parturition starts than later. The signs of impending calving are a relaxation of the muscles around the tailhead, an enlarged vulva and distension of the udder, often with milk leaking onto the ground. The relaxation of the tailhead region occurs about 24 h before the calving and is the best sign of impending calving. The herdsperson may also notice that the cow appears restless and agitated, but this actually starts several weeks before the parturition.

Parturition comprises the following stages:

- Stage 1. Cervical dilation – this stage is terminated by the breaking of the waterbag and appearance of the calf's hooves. The normal duration of stage 1 is 2 h, but it is often longer in heifers.
- Stage 2. Calf expulsion – this normally takes about 1 h. The cow is usually standing initially but lies down to give birth. By contrast, buffaloes give birth in the half-standing position. In this second stage, uterine contractions occur with increasing frequency, usually every 15–20 min, accompanied by strong abdominal straining to contract the diaphragm behind the calf and force it through the birth canal. The expulsion of the head is the most difficult procedure and its size may make it difficult for heifers to achieve calf expulsion unaided. Alternatively, help may be required if the presentation is abnormal, if the head or the back legs are presented first rather than the forelegs. Assistance with difficult calvings should be given initially by the herdsperson, if necessary with the use of traction aids if the cow's efforts to expel the calf seem inadequate. If this is not satisfactory, the veterinarian should be called to give assistance.
- Stage 3. Expulsion of the placenta – this usually takes place 4–6 h postpartum. The cow usually eats the placenta to ward against possible predators

After the calf has been born, the herdsperson should ensure that it is breathing, if necessary clearing the nostrils of mucus and stimulating it by vigorous rubbing and movement. Cow and calf should remain together for at least 24 h to ensure that adequate colostrum has been consumed.

Future Trends

Cattle, both male and female, have over the centuries been revered as a potent symbol of fertility. It is ironic that reproductive difficulties now are responsible for major inefficiencies of production in both dairy and beef herds. The recent decline in dairy cow fertility is particularly of concern, and like the decline in human (male) fertility, is of unknown aetiology. There are many possible reasons – high milk production, intensive and stressful housing conditions, reduced care by the herdsperson or even environmental pollution. It has been demonstrated that one calf per year is attainable over a long lifespan. This is not a high reproductive rate in comparison with other mammals, but when combined with the stress of high production in adverse environments, reproduction is often the first casualty. Difficulty in breeding cattle can only be seen as symptomatic of the poor husbandry of many cattle. Herdspeople are now expected to look after far more cattle than previously, with the aid of mechanization, and this has not assisted cow reproduction. Excessive manipulation of cattle reproduction is likely to meet with public resistance, but the annual production of a calf from each dairy cow may be obviated in the foreseeable future by developments in embryo sexing, by prolonging the lactation and by reduced demand for calves for beef production from the dairy herd. In future it is likely that many calves for beef production will be produced in specialist suckler herds in less favoured regions.

References

Archer, J.A., Richardson, E.C., Herd, R.M. and Arthur, P.F. (1999) Potential for selection to improve efficiency of feed use in beef cattle: a review. *Australian Journal of Agricultural Research* 50, 147–161.

Arthur, P.F. (1995) Double muscling in cattle: a review. *Australian Journal of Agricultural Research* 46, 1493–1515.

Callesen, H., Greve, T. and Avery, B. (1998) Embryo technology in cattle: brief review. *Acta Agriculturae Scandinavica Section A – Animal Science*, 19–29.

Chenoweth, P.J. (1994) Aspects of reproduction in female *Bos indicus* cattle – a review. *Australian Veterinary Journal* 71, 422–426.

Dobson, H. and Kamonpatana, M. (1986) A review of female cattle reproduction with special reference to a comparison between buffalos, cows and zebu. *Journal of Reproduction and Fertility* 77, 1–36.

Emanuelson, U. (1988) Recording of production diseases in cattle and possibilities for genetic improvements – a review. *Livestock Production Science* 20, 89–106.

Essl, A. (1998) Longevity in dairy cattle breeding: a review. *Livestock Production Science* 57, 79–89.

Gordon, I. (1994) *Laboratory Production of Cattle Embryos.* CAB International, Wallingford.

Greve, T. and Madison, V. (1991) In vitro fertilization in cattle – a review. *Reproduction Nutrition Development* 31, 147–157.

Korver, S. (1988) Genetic aspects of feed-intake and feed-efficiency in dairy-cattle – a review. *Livestock Production Science* 20, 1–13.

Odde, K.G. (1990) A review of synchronization of estrus in postpartum cattle. *Journal of Animal Science* 68, 817–830.

Opsomer, G. and deKruif, A. (1999) Post partum anoestrus in dairy cattle – a review. *Tierarztliche Praxis Ausgabe Grobtiere Nutztiere* 27, 39-44.

Phillips, C.J.C. and Schofield, S.A. (1990) The effect of environment and stage of the oestrus cycle on the behaviour of dairy cows. *Applied Animal Behaviour Science* 27, 21–31.

Strandberg, E. (1996) Breeding for longevity in dairy cows. In: Phillips, C.J.C. (ed.) *Progress in Dairy Science.* CAB International, Wallingford, pp. 125-144.

Veerkamp, R.F. (1998) Selection for economic efficiency of dairy cattle using information on live weight and feed intake: a review. *Journal of Dairy Science* 81, 1109–1119.

Further Reading

Broers, P. (Undated) *Compendium of Animal Reproduction.* Intervet International B.V., Boxmeer, The Netherlands.

Felius, M. (1985) *Genus Bos: Cattle Breeds of the World.* MSD Agvet, Merck and Co., Rahway, New Jersey.

Friend, J.B. (1978) *Cattle of the World.* Blandford Press, Poole.

Fries, R. and Ruvinsky, A. (eds) (1999) *The Genetics of Cattle.* CAB International, Wallingford.

Hall, S.J.G. and Clutton-Brock, J (1989) *Two Hundred Years of British Farm Livestock.* British Museum (Natural History), London.

Gordon, I. (1996) *Controlled Reproduction in Cattle and Buffaloes.* CAB International, Wallingford.

Peters, A.R. and Ball, P.J.H. (1996) *Reproduction in Cattle,* 2nd edn. Blackwell Science, Oxford.

Senger, P.L. (1997) *Pathways to Pregnancy and Parturition.* Current Conceptions, Inc., Pullman, Washington State.

Simm, G. (1998) *Genetic Improvement of Cattle and Sheep.* Farming Press, Ipswich.

Stanley, P. (1995) *Robert Bakewell and the Longhorn Bred of Cattle.* Farming Press, Ipswich.

Health and Diseases

6

Introduction

The welfare of cattle principally concerns how they feel – contented, in pain, satiated, excited, to give but a few examples. Their feelings are influenced by whether they have the freedoms that are recommended for any animals managed by humans:

1. Freedom from hunger and thirst.
2. Freedom from discomfort.
3. Freedom from pain, injury and disease.
4. Freedom to express most normal behaviour.
5. Freedom from fear and distress.

Feelings cannot be measured sufficiently well to make deductions about welfare, so biological indicators are normally used. These include behaviour, disease incidence and severity, many physiological indicators, production rate, life expectancy and reproductive rate. Physiological indicators include immune status measures, stress hormones and homeostasis mechanisms, for example pain regulators. These indicators of welfare status are not ideal, in part because they interact, and an indicator that appears to directly increase welfare may decrease it indirectly through its impact on another indicator (Table 6.1). For example, breeding for increased production or reproduction could have negative effects on longevity. This illustrates the need for caution in interpreting measures of cattle production and reproduction to indicate their welfare. There is also difficulty in equating different welfare measures, for example, is an environment that prevents cattle from performing normal behaviour as harmful as one which induces diseases such as lameness or mastitis?

There is a transition from good to bad welfare that suggests that an objective assessment may eventually be possible (Fig. 6.1). A cow in a state of good welfare will be in perfect equilibrium and in a state of contentment; if welfare is reduced she will initially have adverse feelings, then as homeostatic

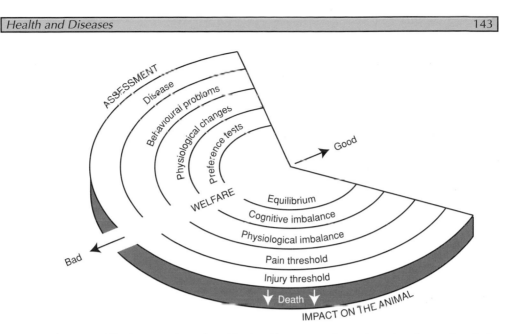

Fig. 6.1. The impact of good and bad welfare on cattle, and means of assessing their welfare state.

Table 6.1. Matrix analysis of the interaction between different indicators of animal welfare (from Phillips, 1997).

	Impact on other welfare indicators[a]					
Welfare indicator	Behaviour	Disease status	Mental satisfaction	Production rate	Longevity	Reproductive rate
Behaviour		N	++	N or +	N or +	N or +
Disease status	+		++	+	+	+
Mental satisfaction	+	+		N or +	N or +	N
Production rate	N or −	N or −	−		−	+, N or −
Longevity	N or −	++	N or +	−		−
Reproductive rate	N	+	−, N or +	+	−	

[a]N indicates a neutral impact on welfare, + indicates positive impact, − indicates a negative impact.

mechanisms break down her physiology will be affected. Pain will be perceived, perhaps in association with, or followed by injury and finally, if corrective action is not forthcoming, death may ensue. The assessment of the cow's state is difficult. Preference tests may inform whether the cow prefers to be in a different state of environment, and her desire for self-administration of analgesics could be investigated to indicate whether she is in pain. As welfare deteriorates, physiological changes become evident, and in particular the stress responses. Behavioural changes may become evident, which when severe can indicate the presence of disease.

The health of cattle is an interaction between the animals, their environment and disease-causing organisms. As they are gregarious animals, cattle should be considered both as a group, but also as individuals if the health of one animal differs markedly from that of the group. Bringing cattle indoors presents one of the most severe challenges to their health as the contact between animals is increased, and disease organisms are better able to survive in the more constant and generally benign environment compared with outdoors. However, nutrition of a high-yielding cow may be better controlled, and the animal's ability to rid itself of pathogens may be better than if they were permanently at pasture, as the immune system functions best in a well-nourished animal. The advent of widespread use of antibiotics in the latter half of the 20th century has heralded a low incidence of infectious diseases for most cattle, especially dairy cows, which are often treated annually at the end of lactation, as well as routinely when there is any evidence of bacterial infection. In the long term, antibiotics will probably have restricted use as the ability of pathogenic bacteria may be greater than our ability to find new antibiotics. Farmers must therefore be prepared to use more prophylactic measures, such as reducing the stocking density of cattle and keeping them cleaner, to prevent disease transfer.

In the last 50 years, farm size has increased considerably in most cattle production systems in industrialized countries, as enterprises expand and small family farms are amalgamated into larger units. Cattle are also more likely to be housed, often to the detriment of their welfare (Hemsworth *et al.*, 1995). The eradication of many disease organisms was achieved in the past through careful husbandry but this could be relaxed with the prospect of effective cures for most of the infectious diseases. The good husbandry that used to be essential for disease eradication is exemplified by the measures taken to prevent the spread of rinderpest more than 150 years ago:

• the byres of infected animals had to be washed and left empty for 2 months;
• persons attending sick animals were prohibited to go near healthy stock;
• the sale of sick cattle was prohibited;
• sick cattle were slaughtered and buried;
• all cases of the disease had to be reported.

These simple measures restricted the transmission of the disease and eventually led to its eradication. Nowadays, there is considerable movement of cattle and this reduces their effective immunity, which develops for a range of pathogens in a specific area.

An alternative method of controlling disease is to reduce the stocking density of housed cattle and hence contact between animals. However, the reduction in output per unit area can only be justified if product prices are increased or farmers are directly compensated for economic losses. Some reduction in cattle stocking density can be achieved through mixed farming systems, for example with sheep, since pathogens find it difficult, but not impossible, to infect stock of more than one species. However, the increased mechanization of farms has necessitated the adoption of simple farming systems, usually relying on the farming of single species of animals and crops (monoculture). It is a technological challenge for the future to design mixed farming systems that minimize disease risks but are still efficient users of land, labour and capital.

The following is not an exhaustive list of cattle diseases, but rather descriptions of some of the most important challenges, in particular mastitis and lameness, and some of the new ones, such as *Escherichia coli* O157 and bovine spongiform encephalopathy (BSE), that are examples of the type of diseases issues that may confront cattle farmers in the future.

Calf Diseases

The newborn calf is relatively unprotected as a result of the *naïveté* of its immune system in responding to environmental challenges. In addition, the calf is growing rapidly compared to its size, so it requires a high level of nutrition. If this is not provided, it will impair the immune system. There is a temptation to reduce costs by limiting milk supply, but calves in the wild or feral calves would naturally suckle their mother for at least 6 months. The health status of many, if not most, calves that are weaned at 6 weeks is generally acknowledged to be worse than that of suckled calves.

During the first 6 weeks, the isolation of calves in individual crates limits cross-infection, but it also prevents the development of normal interactive and locomotive behaviour, and in the long term reduces their ability to socialize with other animals. However, an essential feature of successful calf rearing, which minimizes the disease prevalence, is close observation, and this may be easier with isolated calves than groups. A sick calf is inactive and often lies down for long periods with its head extended. Its eyes may be sunken and lacrimating and its nose and lips inflamed. Its coat is dull and may be soiled with diarrhoea. Stockpeople responsible for calves should be trained to recognize these symptoms, especially when calves are moved away from their place of birth onto other farms.

Calf diarrhoea

Calves suffer from two major forms of diarrhoea or scours – viral diarrhoea, which damages the ability of the intestinal villi to absorb nutrients, and bacterial (usually *E. coli*) or white scours, which does not. It can also result from incorrect feeding of milk powder, either too much in each feed or inadequate quality, particularly protein. Early introduction of solid feed will help to reduce problems with milk feeding. The risk of scours is increased if calves are subjected to stress, for example by movement or following a sudden change of diet.

Calf scours occurs at any time up to 4 weeks of life, by which time the rumen is sufficiently inoculated with benign bacteria to prevent it being colonized by bacterial pathogens. Calves can be vaccinated against certain forms of *E. coli* scours, but the most important means of protection is ensuring that the calf has adequate intake of colostrum in the first 6 h of life.

The major viral pathogens are rotavirus and coronavirus. Calves with viral diarrhoea are not able to reabsorb water in the gut because of villi damage so dehydration is the major problem. Dehydration can be averted by recognizing the symptoms early and providing oral rehydration therapy. It should be accompanied by alkalinizing therapy with sodium bicarbonate as the diarrhoea is usually accompanied by acidosis (Owens *et al.*, 1998), caused by poor renal excretion of hydrogen ions. Reduced nutrient absorption accompanies a severe acidosis and milk should be withdrawn and replaced with a glucose solution for energy.

Calf pneumonia

This infection of the lungs may be caused by several types of viruses, mycoplasma or occasionally bacteria. The conditions in which a calf are kept will determine the impact of the disease, in particular whether there is adequate ventilation and dry conditions, including the calf's bedding, which are essential to minimize the spread and severity of the disease. Traditional, enclosed calf houses with little ventilation and overcrowded conditions are often the cause of an outbreak, whereas modern portal framed buildings with plenty of air for each calf and sufficient air changes to keep the air clean will rarely produce outbreaks. Moving calves from well-ventilated individual pens to group housing at weaning may initiate pneumonia, not least because the challenge faced by the animals when they are grouped together is accompanied by the stress of weaning.

The chief clinical signs are a chesty cough, loss of appetite, sweating and sometimes an eye discharge. An elevated temperature will provide confirmation of the diagnosis. The entire group of cattle should be treated with antibiotics to prevent secondary infections and because inhalation of a small number of bacteria can cause the primary infection.

Other calf diseases

Other common calf diseases include bacterial infections of the navel, especially when the calves are lying on wet ground. These may remain localized in the navel (navel-ill), or circulate around the body and infect other parts, such the leg joints (joint-ill), or creating a serious blood poisoning in the form of septicaemia in the case of *E. coli*. It can also infect the internal organs, such as the liver. Ringworm is an unsightly skin infection of older calves by the fungus *Trichophyton verrucosum,* which can also be transmitted to humans.

Treatment with probiotics

Probiotics are feed supplements that are added to the diet of cattle to improve the intestinal microbial balance. They are more effective in controlling the diseases of the gastrointestinal tract of young calves than older cattle, as there is no complication of the rumen microflora. The colonization of the intestine by benign bacteria may confer protection against pathogenic bacteria. This is not only by competitive exclusion; they can limit the adhesion of some bacteria to the intestinal wall as well and most will improve the immunocompetence of the host animals. Some probiotics, particularly the lactobacilli, can neutralize *E. coli* enterotoxins, and others, notably *Lactobacillus acidophilus*, produce large quantities of lactic acid that reduce pH and prevent the growth of some pH-sensitive bacterial strains.

The initial colonization of the small intestine is from the dam's microflora and the immediate surroundings, and usually includes streptococci, *E. coli* and *Clostridium welchii*. When milk feeding commences the lactobacilli become the predominant bacteria present. Calf probiotics contain benign lactobacilli or streptococci, and are likely to be valuable only when given to calves that have suffered stress or have been treated with antibiotics, which will have destroyed the natural microflora. Addition of probiotics to the diet produces variable benefit, depending on whether the calves are in poor health. It is also difficult to determine which bacterial species would be beneficial in each circumstance.

Cattle Lameness

Lameness is characterized by a departure from the normal gait, caused by injury or disease in a part of the limbs or trunk, and usually accompanied by pain. It is a problem that is increasing in severity in many countries, and in the UK it currently affects at least 20% of dairy cows each year (Politiek *et al.*, 1986). It is particularly associated with high yielding cows in cubicle housing. Fifty years ago, when most cows in the UK were individually housed in byres and fed a hay-based diet with only a small amount of concentrate, the incidence was less than 5% of cows per year. However, the increase in lameness over time is difficult to estimate precisely as different recording methods are used. Early studies relied mostly on veterinary records, but recent studies take into account the records of foot trimmers, the herdsperson and also sometimes the locomotion scores of the cattle. A five-point score is most commonly used: 1 – perfect locomotion; 2 – some abnormality of gait; 3 – slight lameness, not affecting behaviour; 4 – obvious lameness, behaviour affected, and 5 – difficulty in getting up and walking. Some recent UK surveys have indicated that the incidence is over 50%, and we can only conclude that lameness is now an extremely serious problem for the dairy industry in the UK.

Most lameness occurs in the hind feet, especially the outer claws. It is actually a number of different disorders, in particular sole ulcers, white line lesions, laminitis, digital dermatitis and interdigital infections, each of which is briefly described.

Sole ulcer

A sole ulcer is manifested as a haemorrhage in the sole, most often in the lateral claw of the hind feet. This is because of the pinching of the corium by pressure from the pedal bone, which provides an entry route for bacteria (Fig. 6.2). Often the ulceration is hidden by a thin sliver of horn tissue, but paring this away reveals the haemorrhaging. Sole ulcers are one of the most common forms of lesion. Predisposing factors include lack of exercise, subclinical laminitis and wet conditions.

White line lesion

The white line is where the horn of the hoof wall meets the sole. It is a junction cemented by immature, unpigmented horn tissue, which is weaker than older, more highly keratinized horn tissue, and is therefore prone to entry of foreign bodies. These may be forced upwards to the corium by the pressure of the cow walking on an injured site, where an abscess forms. Pus may accumulate within the hoof and has to be drained out. When cows walk on stony ground, foreign bodies can puncture the sole and cause pus formation, which has to be drained out by a vet or professional foot trimmer. White line lesions are particularly common in wet conditions.

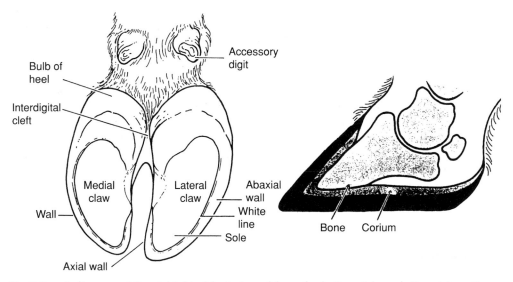

Fig. 6.2. A diagram of the right, hind foot viewed from the bottom (above left) and from the side (above right). Reproduced courtesy of Farming Press Books Ltd.

Laminitis

This is an inflammation of the laminae of horn tissue that are produced from the modified corium or papillae just below the coronary band. The inflammation is a common condition in lactating cows housed on concrete and fed silage and concentrates, but the aetiology is not yet well understood. The principle influence of concrete is the high impact that it inflicts on the hoof, causing a sinking of the bone illustrated in Fig. 6.2 (the pedal bone) within the claw capsule and putting pressure on the corium. However, equally important may be the release of substances (perhaps bacterial endotoxins) in the corium that cause vasodilation and anoxia, leading to the failure of the laminae to support the weight of the animal. This is probably a consequence of the feeding regimes that cows are given in modern dairying systems. Excessive concentrate feeding causes a low and widely fluctuating pH of rumen fluid, which may induce such conditions in the hoof. Whatever the reason, or reasons, for this disease, it causes painful haemorrhaging in the hoof laminae and predisposes the hoof to other disorders, leading to a possibility of a common aetiology of several hoof disorders, arising from the housing and feeding system currently used for most high-producing dairy cows.

Digital dermatitis and related diseases

This is a newly recognized disease that has rapidly become one of the most common causes of lameness. It is associated with housed cows and exacerbated by wet conditions. The symptom is a painful skin lesion, usually at the back of the foot. It appears to be caused by bacteria, as it will respond readily to antibiotics and is highly contagious if not treated. Probably *Bacteriodes* species or spirochaetes are responsible. Footbaths are sometimes used for treatment, but in the UK, there are currently no antibiotics licensed for use in baths, and antimicrobials are rapidly neutralized by the large amount of organic matter in the footbath.

Many regard interdigital dermatitis as of the same aetiology as digital dermatitis, since the most commonly isolated organism is *Bacteriodes nodosus*. The same is true for slurry heel, although laminitis is a strong predisposing factor in this disease. The erosion of the heel bulb caused by the bacteria is often prevalent throughout the older animals in a herd. It does not itself cause lameness but predisposes to other hoof disorders, and may be considered as part of the general hoof condition caused by the management system for high-yielding, housed dairy cows.

Interdigital necrobacillosis

Otherwise known as foul-in-the-foot, this disease is caused by a bacterial infection, *Fusiformis necrophorum*, that commonly inhabits the environment of cattle and sheep and can survive in the soil for 10 months. It cannot penetrate the interdigital skin tissue, unless there is damage by stones, sticks, etc., when it invades the underlying tissues and causes skin lesions. It causes a marked swelling and acute lameness but is not fatal. Regular cleaning of the hooves by walking cows through a formalin footbath will help to prevent this disease.

Housing, cow tracks and leg disorders

The type of housing is the dominant influence on the incidence of the common disorders of the legs of cattle. Concrete wears the hooves more than earth, but the constant wetting of the foot in cubicle passageways covered in deep slurry can erode the soft heel bulb. This predisposes the cow to laminitis and sometimes leads to the toes losing contact with the ground and unchecked growth (slipper foot). Excessive walking on concrete stretches the white line and wears the sole, thus weakening the bond between the wall and the sole. In straw yards the abrasion on the hoof is minimal, leading to increases in toe length until the cows are out at pasture. Interdigital infection is more common than in cubicle houses as straw may be pushed up between the claws, causing a lesion that is open to infection, especially with *Phlegmona interdigitalis*. Excessive growth may close up the interdigital space, trapping dirt and causing infection. Long digits are common in modern cattle units partly because of excessive growth as a result of the high nutrient density of their diet and partly inadequate wear as a result of low levels of activity and a smooth concrete floor or soft strawed yard to walk on.

Slipping can be a significant problem in cattle accommodation and slaughterhouses, especially around water troughs, and where cattle are moved rapidly and are turning sharp corners. The hooves may be trimmed to correct the hoof conformation, which will increase the area of the hoof in contact with the floor and reduce the chance of a cow slipping. Trimming reduces the load bearing of the heel, improves locomotion in cattle and reduces the hardness of the sole and abaxial wall. It is normally done by the following method, first developed by Dutch cattlemen:

1. A cut to the wall of the medial claw is made at the toe-end with hoof clippers, perpendicular to the sole and 75 mm from the periople line. The sole is then pared with a knife to remove any overgrowth, reducing the original cut to the digit to 7 mm and exposing the white line.
2. Using the medial claw as a template the lateral claw is cut to the correct length and the sole pared to the same depth.
3. The non-weight-bearing axial surfaces of the sole of both digits (the area surrounding the interdigital cleft) is hollowed out.
4. Irregularities in the sole are excavated and corrected.
5. Heels are trimmed to remove loose horn and reduce any furrows.
6. The soles are levelled off using a rasp.

The condition of cow tracks can influence the prevalence of lameness in a herd. Stony, muddy tracks provide an uneven surface that can stress the sole and lead to lameness and poor welfare. The ideal surface is absorbent and provides a firm surface for walking on — bark chippings are recommended, but need constant care. They should be laid over a porous membrane and aggregate for good drainage. Concrete is very durable and is better than stony tracks, but not as comfortable for the cows as bark chippings. However, a concrete track can be used by both cows and vehicles, whereas if bark chippings are used separate tracks are required as a vehicle's wheels will create ruts that would make comfortable walking difficult for the cows. Cows should never be hurried down a track by a herdsperson using a dog or motorbike.

Mastitis

Mastitis is an opportunistic infection of the mammary gland, the severity of which is mainly dependent on the delay in inflammatory response to the infection. It is not one disease, but a wide range of possible infections by up to 100 possible pathogens, making it a multifactorial disorder.

Mastitis causes a major reduction in the welfare of the cow, because of the fever that it induces, the localized pain in the mammary gland and the possible stress of isolation and treatment. The cost to the farmer is considerable, with losses arising for a number of reasons:

- penalties imposed by the dairy purchasing the milk because it has a high somatic cell count (SCC);
- the reduction in milk yield from the infected gland, which may persist into the subsequent lactation;
- the withholding of the milk from sale after antibiotics have been applied;
- drug and veterinary costs;
- reduced value of the affected cow;
- replacement costs;
- increased labour for managing the sick cow;
- the cost of associated diseases, in particular reproductive failure.

Mastitis also presents risks to human health, as milk containing antibiotics that is drunk by humans may induce them to develop bacterial resistance, or the consumption of unpasteurized milk contaminated with bacteria could lead to the consumer acquiring zoonotic infection. There are penalties in most countries for dairy farmers supplying milk contaminated with antibiotics, and insurance against this possibility is now available. Not only is milk contaminated with antibiotics unfit for human consumption, it is also unsuitable for cheese or yoghurt production, as the antibiotics prevent the fermentation proceeding. Dairy cows that contract mastitis before pregnancy is established have a delay in the interval to first service, on average about 3 weeks, and an average of one extra service is required for each conception.

Mastitis signs

Chronic infections are most common, affecting 20–25% of cows annually in the UK. It is useful to recognize these cows in the early stages of infection, so that they can be milked after the cows that are free of infection. This will minimize transfer between cows. In chronic mastitis, some abnormal secretions are usually evident, but in acute cases the gland is swollen, hot and full of clotted milk and the cow suffers from a potentially fatal fever. Dairy cows on organic farms have a particularly high incidence of subclinical mastitis, but many of these recover spontaneously. Mastitis can be detected by foremilking, which is a legal requirement in Britain, by in-line mastitis detectors, and by palpating the udder and detecting its hot, swollen appearance.

Defence mechanisms

The external surface of the mammary gland is designed to minimize the adherence of bacteria, being without hair or sweat and sebaceous glands. The teat is the primary line of defence against mastitis, with its skin possessing a thick layer of stratified squamous epithelial tissue. The surface contains much dead keratinized tissue that is a hostile medium for bacterial growth. At the end of the teat is a 8–12 mm canal surrounded by a sphincter that closes within 20–30 min of milking because of pressure of milk on the Furstenburg's rosette (a ring of muscle around the top of the canal). This prevents the ingression of bacteria into the teat cistern. It is therefore wise to prevent housed cows from lying down for 30 min after milking, which is usually achieved by restricting them to the feeding area. Usually cows are fed in the morning but if possible the feed should be offered at the feeding barrier during afternoon milking so that cows spend time feeding on returning to their housing.

Some of the keratinized epithelial tissue lining the cistern is sloughed off during milking, preventing adherence of foreign material to the gland. Bacteria may contaminate the teat end from the milker's hands, the environment (particularly bedding material), udder cloths and residual milk from the previous cow to be milked. The keratin itself has antibacterial properties, being a waxy substance that limits moisture availability to colonizing bacteria. In high-yielding cows, there can be excessive shedding of keratinized epithelial tissue into the teat canal, which reduces its protective properties between milkings.

Other non-specific defence mechanisms include the lactoperoxidase complex in milk (lactoperoxidase, thiocyanate and hydrogen peroxidase), which is bactericidal to Gram-negative bacteria such as *Streptococcus dysgalactiae* and can inhibit the growth of Gram-positive bacteria. Lactoferrin in milk provides a second biochemical defence mechanism. It is an iron chelator that depletes milk of the available iron that is required by bacteria. It can be inhibited by the presence of citrate during lactation. Similarly, the increase in milk sodium during mastitis is probably toxic to some mastitogenic bacteria, particularly

Staphylococcus aureus. Also in the milk are complement proteins and immunoglobulins, both of which are bacteriostatic or bactericidal.

The specific defence systems that are used to protect against bacterial colonization of the gland centre on the polymorphonuclear leucocyte (PMN) response to macrophage recognition and destruction of invading bacteria. This inflammatory response can increase cell counts very rapidly, because of the logarithmic increase of PMN numbers when lysosomal breakdown products are liberated in the gland. Of particular importance is the rate of arrival of neutrophils at the mammary gland, which is often 2–4 h after the appearance of endotoxins in the gland. Neutrophils in the milk have less phagocytic activity than blood neutrophils. Adequate zinc, selenium and vitamin E nutrition has been shown to be particularly important for neutrophil function around calving and early lactation. During the infection, the paracellular junctions in the secretory tissue become more permeable, and sodium leaks into the milk, giving it an unpleasant taste. This can be detected by the change in electrical conductivity of the milk, providing a parlour test for both clinical and subclinical mastitis detection. However, even if detection can be improved for subclinically infected cows, therapeutic measures are restricted to limiting the possibilities of transfer to uninfected cows. This can be done, for example, by milking the infected cows last in the herd. Routine antibiotic use on subclinically infected cows would not be advisable because of the risk of antibiotic resistance developing, especially since many subclinically infected cows recover spontaneously.

Somatic cell counts and other milk quality measures

The somatic, or body, cells are mostly macrophages, which reside in the teat cistern. The somatic cells in normal milk produced in mid-lactation typically comprise 65% vacuolated macrophages, 14% non-vacuolated macrophages, 16% lymphocytes, 3% PMNs and 2% duct cells. Macrophages recognize invading bacteria and invoke the production of PMNs to phagocytose invading bacteria. The duct cells are epithelial and alveolar cells that apoptose and are excreted in milk. These represent the basal level of somatic cells in milk (the somatic cell count, SCC).

There are three main disadvantages to high milk SCCs. Firstly, milk yield is reduced because of damage to the secretory tissue, by about 2.5%/100,000 cells ml^{-1} above 200,000 cells ml^{-1} (Blowey and Edmondson, 1995). Secondly, the milk has increased lipase content, and the lipid breakdown products give it a rancid taste. Thirdly, the milk has a low casein content leading to reduced cheese yield. In 1992, a legal limit of SCC in cow's milk for human consumption of 400,000 cells ml^{-1} was included in the EU Hygiene Directive, but the mean SCC in many countries is well below 250,000 cells ml^{-1}. At a herd level or over a long period of time, the SCC is an indicator of the level of bacterial infection in the mammary gland. However, for an individual cow, an increase in SCC may not coincide with an acute bacterial infection, since the SCC can take 2 weeks to decline after treatment (Laevens *et al.*, 1998). The national level of SCC has decreased considerably since incentive payments and penalty schemes were introduced. However, at the same time milk yield increased, so some of the decrease is caused by dilution. Mastitis incidence has not declined, and remains approximately 35–40 cases per 100 cows per year in Britain, so it is likely that most of the decrease in SCC is because of farmers being more aware of the need to prevent milk from cows with a mastitic infection entering the saleable milk in the bulk tank.

Cows with no signs of clinical mastitis produce milk with an SCC of about 150,000 cells ml^{-1} on average in their first lactation. In subsequent lactations, SCC usually increases, and is likely to be up to 300,000 cells ml^{-1} (Laevens *et al.*, 1998). Since milk can only be purchased from an EU farm if the SCC is less than 400,000 cells ml^{-1}, it is imperative that the incidence of mastitis be kept to a minimum.

The SCC only increases when the udder's defences have failed to prevent bacterial colonization of a gland. The California Mastitis test gives an approximate guide to the SCC, by testing for the amount of DNA in a milk sample. The CMT values range from 0 (< 200,000 SCC) to 3 (SCC > 5,000,000). It can be a good screening test for large numbers of cows in a problem herd, and any samples with test values of one or more indicates that a more detailed bacteriology test should be conducted.

In addition to SCC, there are direct measures of bacterial contamination of the milk that are routinely conducted. The total bacterial count (TBC), which relies on culture to determine the number of bacteria in a sample, has been replaced in Britain by a Bactoscan measurement, which dyes and counts bacteria automatically. The values do not correlate directly with TBC values, because only the Bactoscan measures psychotrophs (bacteria capable of growing in cold conditions), but generally the Bactoscan measurement is 4–5 times the TBC measurement. The TBC or Bactoscan measurement not only increases with mastitis; inadequate refrigeration and dirty milking procedures or cows' teats may also be implicated in high values. Some of these contributors to a high TBC level can be detected directly, if it is suspected that mastitis is not the cause of the problem. For example, a coliform count of more than 20–25 ml^{-1} indicates that the cows have dirty teats or that cows are being washed but not dried, which conveys the bacteria into the liner and then into the milk. A high number of thermoduric bacteria (indicated by a laboratory pasteurized count of more than 175 ml^{-1}) indicates inadequate washing, since thermoduric bacteria multiply in a film of milk that remains in the milking machinery after inadequate washing of the plant. This could be because of a boiler failure, insufficient washing solution or blockages in the washing system.

The pathogens causing mastitis

The mastitogenic pathogens can be divided into those that are contagious and are passed from cow to cow, often via the milking cluster, and those that are contracted from the environment, particularly dirty, moist bedding. Over the last 30 years or so, there has been a reduction in contagious pathogens because of the widespread use of prophylactic measures, while environmental mastitis forms have increased. The most common contagious organism is *Staph. aureus*, which is usually spread from cow to cow at milking through the milking machine or by the milker, and is persistent once established. The most effective control is good parlour hygiene and dry cow therapy. *Staph. aureus* bacteria produce toxins that bind to the epithelial membrane of the secretory tissue, opening transmembrane pores that cause the leakage of ions from blood into the milk. Eventually secretory tissue is destroyed and there is an accumulation of fibrous tissue that inhibits antibiotic activity. *Staph. aureus* bacteria are particularly resistant to antibiotic therapy, in part because they are attached to milk fat globules, which encapsulate and protect them. One of the control measures against mastitis is to regularly extract as much milk as possible, perhaps by hand stripping or allowing a calf to suckle. This helps to extract the fat globules, which are not easily expressed from the milk tubules into the collecting ducts. At the end of lactation, infrequent milking may lead to pathogens proliferating rapidly, in which case the lactation should be discontinued.

Other contagious mastitogenic pathogens include *Strep. dysgalactiae*, which can exist in extramammary reservoirs, most notably the tonsils, *Streptococcus agalactiae*, which exists in bedding and milking equipment and causes large increases in bulk tank SCC, mycoplasma and *Corynebacterium bovis*. *Strep. agalactiae* is spread easily between cows at milking time. However, it only survives in the mammary gland and is susceptible to anti- biotics. *C. bovis* dramatically increases SCCs but often with limited infection. It exists mainly in the mammary gland and reproductive tract.

The major environmental pathogens are *E. coli*, which comes mainly from cattle faeces because of its presence in the intestinal tract, *Streptococcus uberis*, from external surfaces of the cow and some orifices and which also causes large increases in milk SCC, and *Klebsiella* spp., which are particularly common in damp sawdust. *E. coli* infection is particularly prevalent at the onset of lactation in high-yielding cows and shortly after the cessation of lactation. It is much more pathogenic than *Strep. uberis*, producing toxins that rapidly induce loss of appetite, fever, depression and a reduction in milk production of about 60% within 2 days. It is the most likely of the mastitogenic pathogens to cause an acute, and sometimes fatal, mastitis.

Mastitis is generally most prevalent in the immediate postpartum period because of the suppression of immunocompetence which lasts for about 1–2 weeks before and after calving. This periparturient immunosuppression is believed to be partly because of the preparation for and onset of lactation, partly because of changes in the environment and greater bacterial challenge in early lactation, and probably partly because of the rapidly escalating nutritional deficit at this time.

Another important mastitogenic pathogen is *Actinomyces pyogenes*, which is both infectious and of environmental origin and resistant to antibiotics. This is the main cause of summer mastitis that infects dry cows and heifers, but rarely lactating cows; it is transmitted by the headfly, *Hydrotea irritans* and is best controlled by fly elimination with impregnated ear tags or sprays (Byford *et al.*, 1992).

Surveys indicate that on a national basis the proportion of mastitis incidents are distributed amongst the bacteria as follows: *E. coli* 27%, *Strep. uberis* 21%, *Staph. aureus* 19%, *Strep. dysgalactiae* 6%, *Strep. agalactiae* 2% and other bacteria 25%. Previously, *Staph. aureus* was a more common cause of mastitis, but the widespread use of intramammary antibiotics and better milking hygiene has reduced its prevalence, with *E. coli* increasing in its place.

Control measures

Control measures are based mainly on cleanliness and the regular use of antimicrobial chemicals and antibiotic drugs. They have been well established for at least 25 years and, if correctly implemented, should enable a dairy herd to have between 100,000 and 150,000 somatic cells per ml of milk in the bulk tank, fewer than 20 cases of mastitis per 100 cows in a 305-day lactation, less than 12% of the herd infected at any one time and a recurrence rate of less than 6%. Less than five doses of antibiotics should be used for each case of clinical mastitis. These targets, although attainable, may not be adequate for milk purchasers in future, who may devise species specific penalties/incentive payments to regulate individual bacteria.

The successful treatment of mastitis will require researchers to continue to develop better control measures, since the bacteria will mutate and develop resistance to the current range of antibiotics. So far, resistance has only been observed in response to *Staph. aureus* infections when these are treated with penicillin or when cloxacillin is used for dry cow therapy. *Staph. aureus* infections during lactation are best controlled by clavulanate/amoxycillin, and for dry cow infections, cloxacillin is still the best antibiotic to use. Most other infections can be controlled with penicillin, although clavulanate/amoxycillin is best for *E. coli* infections. When choosing an antibiotic, careful note should be made of the milk-withdrawal time (from the time of insertion of the last tube). Some products are a combination of antibiotics. When using antibiotic therapy on dry cows, the persistence should be noted, which ranges from 7 days for ampicillin to 42 days for cloxacillin.

The level of mastitis has not declined significantly over the last 20 years in the UK despite widespread adoption of control measures, and in future farmers will have to concentrate on maintaining their cows in very clean conditions. This will be particularly relevant in controlling environmental pathogens, which have become more common as infectious pathogens have been brought under control by antibiotics. Slurry disposal will need to be carefully planned and attention paid to fly control and ventilation of buildings.

Heifers are particularly vulnerable to mastitis as their defences are not well developed, and they can have twice the incidence of mastitis found in older cows. Infection chains from older cows to the heifers are common, where bacterial infections are transmitted in the milking parlour.

Part of the reason why mastitis incidence has not declined relates to the increase in milk yield of cows in most areas. High yielding cows tend to have wide teat canals, so the ingress of bacteria into the teat cistern is facilitated. Increasing milk yields through the injection of the growth hormone analogue (bovine somatotrophin or bST) also increases mastitis. The bST can stimulate the immune system in the mammary gland, but the extra mobilization of body tissues to support increased milk production may generally depress the immune function, leading to an increased mastitis frequency.

The current recommendations for controlling mastitis are presented in a five-point plan (Box 6.1), but these may have to be modified as antibiotic resistance reduces the effectiveness of antibiotics in combating new strains of bacteria.

Box 6.1. The five C's for control of mastitis.

---

1. Clinical cases treated.
Treat all clinical cases promptly with the vet's recommended intramammary antibiotic and record the cases accurately.
2. Cows treated at drying off.
Treat all cows at drying off with a long-lasting dry cow intramammary antibiotic, applied to each quarter.
3. Clean cows.
Use a teat dip or spray to disinfect all teats after every milking. Clean cows' teats before milking with hot water and dry with individual paper towels.
4. Clean machinery.
Ensure that the milking machine is tested regularly and faults promptly corrected. Maintain the parlour in a clean state and clean the equipment thoroughly on a daily basis.
5. Cull persistently infected cows.
Cull animals with persistent and recurring cases. Break the infection chains.

Some explanatory notes:

Clinical case treatment

Treatment should be given to the cow according to the severity of the disease. Milk cases should be treated with intramammary tubes for 3 days, administered by the farmer. If mastitis is severe, the vet may be needed to administer high-dose injectable antibiotics, as well as intramammary treatment. Hypertonic saline solution fluid therapy may be given if the case is particularly severe. This should be given intravenously and will be beneficial in counteracting the dehydration that accompanies acute mastitis.

Cows treated at drying off

Some bacteria, such as *Staph. aureus* and *Streptococcus* may be active from one lactation to the next, so the routine use of an antibiotic intramammary infusion at drying off will prevent recurrent inflammation of the gland. The cure rate will depend on the level and duration of the infection, the number of quarters that are infected and the age of the cow. Older cows are much more difficult to cure, partly because they develop resistance to antibiotics, in particular penicillin. Cure rates vary from about 90% for a 3-year-old cow with one infected quarter to about 35% for an 8-year-old cow with three infected quarters. The spontaneous cure rate is normally 10–15%.

The routine use of antibiotics for the control of contagious mastitogenic pathogens, particularly at drying off, is probably partly responsible for the recent increase in environmental pathogens. As the contagious pathogens are treated and eliminated, the mammary gland has been left susceptible to novel pathogens, in particular those normally present in the environment. In the long term, alternative treatment methods will have to be found, and more emphasis focused on prevention.

Clean cows

Dirty bedding contains *Strep. uberis, E. coli* and other coliforms. Outside areas where the cows lie overnight can become soiled and harbour *Strep. uberis*. In the parlour, all cows should be washed on entry to their stall if their teats are visibly dirty and then dried with individual paper towels. A communal cloth spreads infection from cow to cow. If the teats are not visibly dirty, they should be wiped with a dry paper towel.

Cows teats may be either sprayed or dipped in a disinfecting solution, usually based on iodine, together with an emollient to stop the skin dehydrating and chapping. These reduce the rate of new infections by *Staph. aureus* and *Strep. agalactiae* by about 50%, but are ineffective against *E. coli*. Alternatively, hypochlorite solutions are effective but sometimes irritate the skin. Spraying is quicker but dipping ensures a better coverage of the teat. Before milking, the udder should be washed with a sprayline and individual paper towels used to dry the udder, with the milker wearing disposable gloves to prevent bacteria on his or her hands contacting the udder. This will help to limit the invasion of the udder by contagious pathogens, whereas post-milking teat-dipping largely controls the environmental pathogens.

Clean machinery

Machine maintenance is essential at least once a year to ensure that the teats are not excessively stressed by the milking machine. This will always be difficult to achieve as the sucking action of the machine inevitably puts more pressure on the teat blood vessels than the calf, which mainly squeezes the milk out of the teat with its tongue. Rapid build-up to a maximum vacuum and return to atmospheric pressure are important. The vacuum should be maintained for at least one-third of the cycle and fluctuation in pressure within the open phase avoided. Valves can be inserted into the claw to prevent transfer of milk from one teat to the other. Liners should be designed to minimize slip and should be replaced every 10,000 milkings or 6 months to prevent bacteria being harboured in damaged liners (Bramley, 1992).

Culling persistently infected cows

The low limit for SCC in milk in the EU should encourage farmers to regularly cull cows with high SCCs. As the SCC increases with lactation number, older cows will naturally be a target. Farmers should remember that SCCs are lower in mid-lactation, when yields are greatest, than at the beginning or end of the lactation. However, it may be unwise to breed cows selectively with very low SCCs, since these cows may be more at risk of developing mastitis, particularly from the environmental organisms such as *E. coli* or *Strep. uberis*. The colonization of a gland by minor pathogens, which increase SCC but do not cause clinical mastitis, reduces the risk of contracting a severe environmental mastitis. Some argue that the speed of reaction of the immune system is more important, but currently this cannot be included into any widespread breeding programme.

Future control measures

In future control will have to rely to a much greater extent on prophylactic measures than the use of antibiotics. The following are some of the measures that farmers will have to focus on:

1. Minimizing contact between the cow and slurry/faeces.
2. Avoiding muddy pastures and cowtracks.
3. Culling cows that lie in passageways rather than cubicles or strawed yards.
4. Improving cubicle design to encourage cows to use them.
5. Training cows to use cubicles.
6. Cleaning cubicles regularly and improving bedding provision to reduce bacterial contamination of the bed.
7. Cleaning the cluster between cows during milking.

Bovine Spongiform Encephalopathy (BSE)

On 22 December 1984, the first believed case of BSE was seen in Sussex, UK. The animal died 3 months later but was misdiagnosed as having worms or mercury poisoning. In 1986, the disease was eventually confirmed as a spongiform encephalopathy by the UK Ministry of Agriculture, Fisheries and Food (MAFF), but the diagnosis was suppressed for 18 months. It was originally believed to be one of a few isolated incidents caused by toxic material; however, by the end of 1986 there were seven confirmed cases. The new disease was a fatal, transmissible, neurological condition, with the infective material being detected in the brain, retina, spinal chord, ileum, some ganglia and the bone marrow of affected animals.

The confidence that Europe had in its cattle industry was then shattered in 1987 when it became apparent that the UK had become the centre of a major new epidemic – a transmissible spongiform encephalopathy (TSE) (Donnelly *et al.*, 1999). TSEs were known in other species but there was no known cure or even preventative measure, other than the certain knowledge that there was a strong environmental component. Slaughtering sheep flocks with the ovine TSE, scrapie, and restocking had not eliminated the disease, which has remained at a low level in certain sheep flocks for many years. Because of this, the suspicion arose that the recycling of meat and bone meal from sheep had caused the TSE to transfer species into cattle. This association was strengthened by the observation that the brain tissue of cattle with BSE contained scrapie-associated fibrils made of prion protein. In particular, it was suspected that infective agent had been delivered to farms because of the sudden and widespread appearance of the disease (it is now known that sheep too develop scrapie through infection, not as a spontaneously generated disease). It transpired that the rendering industry had recently revised its practices, under the guidance of the UK Ministry of Agriculture, Fisheries and Food, in order to reduce energy costs and reduce the risks to abattoir workers through the removal of acetone from the process of fat extraction from the carcass and a reduction in the temperature at which the extraction took place. Although there have been anomalous situations where cattle that have contracted BSE appear not to have been fed meat and bone, it now seems very likely that this was indeed the true origin of the disease, because the prion glycoforms were identical for the two diseases. Most cattle were probably infected as calves in 1981/82. Another change in the cattle industry that took place at this time was the widespread use of systemically active organophosphates that were poured onto the backs of cattle to eradicate warble fly, which is suspected of predisposing cattle to the transmission of the disease from sheep.

In 1988 the UK Ministry of Agriculture, Fisheries and Food made BSE a notifiable disease and prohibited the feeding of any ruminant protein to ruminants. It is now questionable whether this was a sufficient reaction to the threat of widespread transmission of the disease, because it was known at the time that a wide variety of animals could become infected with TSEs from other species, albeit less easily than the transmission of infection within a species. Meat and bone meal had been used in the cattle feed industry for about 40 years, and in the late 1980s, about 13,000 tonnes of meat meal and 5000 of bone meal were exported annually from the UK by the rendering industry. As a result, the disease has made sporadic appearances in other countries, but the reporting of these has been limited because of the obvious damage that such cases can do to a country's cattle industry. Nowhere has that damage been more acutely felt than in the UK. In fact, it is no exaggeration to say that the disease threatened the survival of the government in power at the time, who had to negotiate a delicate path between on the one hand over-reaction and destruction of the UK cattle industry and on the other taking adequate steps to safeguard the UK population from acquiring the disease. Repeatedly the government of the day tried to allay public suspicions by claiming that there was no risk associated with eating beef. When it became apparent that public suspicion was not satisfied by these pronouncements, the government authorized, in 1989, the removal of specified offal and nervous tissue in abattoirs, and these items were effectively removed from the human food chain. Repeatedly, several senior scientists warned that this was not enough, and in the same year the EU voted to ban UK exports of beef, believing that some infective material was still reaching the human food chain. The offal ban was extended to intestines and thymus in 1994, after infective agents were found there in calves after oral infection. A selective cull was started in 1996 with over 60,000 animals over 30 months of age being slaughtered out of total of 160,000 cattle that had been believed to be infected, an operation which was far more expensive than if the government had introduced greater controls at the start of the crisis, because of some vertical transmission of cases and spread of the disease to the majority of farms in the UK.

In the same year a new variant of Creutzfeldt–Jacob disease (nv-CJD; a rare spongiform encephalopathy affecting the human population) was described in the UK, and several scientists linked its emergence to the BSE epidemic when similar patterns of glycosolation and behaviour were observed in mouse bioassays for the two diseases. Since 1996, 10–20 cases of nv-CJD have been identified each year, although it is currently unclear whether these people had particularly short incubation periods, were more susceptible or consumed a greater infective dose than the rest of the population. The increasing suspicion that the disease could transmit to humans led to ever more stringent measures to try to control its spread. In 1988, the UK government banned the use of ruminant protein in ruminant diets, in 1994 this was extended to all mammalian protein and in 1996 the ban applied to all farm animals, not just ruminants. Over this period of time, it was seen that BSE could be experimentally transmitted to many other mammals by injection into the brain and to some mammals (other cattle, sheep, goats, mink and mice) by oral ingestion of infective agents. As little as 1 g of infected bovine brain material could cause development of the disease via the oral route, and considerably less in sheep and goats. Natural transmission occurred to domestic cats, captive wild ruminants and carnivores. As the epidemic lasted from 1987 to 1998, many of these measures can be seen, with the benefit of hindsight, to have been too little, too late.

If there is one lesson from the BSE outbreak in the UK during the 1980s and 90s for those involved in the cattle industry, it is that disease outbreaks are a constant threat to their livelihood. At any time, a major disease outbreak can leave a cattle farmer unsupported by government, the enemy of the public and without a market for their product. The impact on the livestock industry in the UK has already been very significant, but it remains to be seen whether it can recover and what the zoonotic effect of the disease is.

E. coli O157

E. coli O157 is an example of an evolving new bovine pathogen, which emerged at the beginning of the 1980s. Its effect in humans is to produce toxins that attack the small blood vessels in the kidney, brain or large intestine. Renal failure is the most common cause of death. The organism is resident in the faeces of some cattle and, although it does not cause any disease in ruminants, it is a potent zoonotic agent that is transmitted in meat and milk through contamination of the coat and udder with faeces. In liquid milk it is transmitted only if pasteurizing processes are absent or inadequate, but it can also be transmitted in milk products such as cheese.

Ensuring that cattle are kept clean and clipping off contaminated hair before they are sent to slaughter will limit the spread of the organism to humans. A dirty animal may have 10–12 kg of manure on them, which reduces the value of the pelt as well as creating a health risk to people handling and consuming the meat of the animal. Dirty animals should be rejected at the abattoir and returned to the farmer. Clipping to remove faeces from the pelt may present a risk to people doing the clipping unless adequate precautions are taken. Prevention can also be achieved by better microbiological training of food preparation staff, including farmers with milk-processing plants on their farm. Part of the reason for the organism's success is its ability to survive in harsh conditions. Laboratory tests show that it can survive for several days on dry stainless steel. It also survives in soil, from which it can cross-contaminate stock that are lying down outside.

The worst outbreak so far occurred towards the end of the 20th century in Lanarkshire, Scotland, when 21 people died as a result of having eaten steak and kidney pies produced from a contaminated side of beef. The speed with which the organism spread is a major concern in attempting to guard against future outbreaks. In this case the authorities acted quite quickly and closed the shop selling the infected meat within 10 days of it entering the premises. Inadequate hygiene and disinfection procedures had allowed the organism to spread throughout the premises. Rapid recognition of the problem and treatment of infected individuals is essential in such outbreaks, and inevitably it is the old and infirm that are most likely to die. Nevertheless, the public tends to see the risk as disproportionately high, because it is outside their control. The peak incidence in Scotland, which has a greater prevalence of the disease than any other country in the world, was 10 cases per 100,000 in 1987, which is a small risk compared to other possible causes of death.

Bovine Tuberculosis

This disease usually causes localized infections (tubercles) in lymph nodes, especially in the respiratory tract, although it can spread to other parts of the body, such as the mammary gland, if it is undetected (Krebs *et al.*, 1997). It is occasionally found in the gastrointestinal lymph nodes, but the infective dose for gastrointestinal infection is much greater than for respiratory infection, where a single organism delivered to the right place may suffice. The responsible organism is *M. bovis*, which used to be responsible for many thousands of deaths in the UK, when children in particular drank the milk of infected cattle. In the 20th century, the widespread adoption of pasteurization reduced the level of infection in humans to those that directly came into contact with infected cattle. In addition, a programme of tuberculin testing cattle regularly for *M. bovis* reactivity and slaughtering positive reactors was begun in the 1930s. This was initially on a voluntary basis, but in 1950, a compulsory programme was introduced that has continued to this day. This reduced the number of breakdowns to less than 0.1% of UK herds by 1970. However, it is a good example of the difficulty encountered in completely eradicating a disease organism. The disease has been increasing again since the late 1970s, and the number of UK herd breakdowns increased nearly fivefold in the 1990s to approximately the same level as in 1948. This may be the result of the growth in population of an intermediary host, the badger, which used to be controlled by game wardens. The increased use of maize silage and complete diets on dairy farms may also have increased the population. Cattle probably contract the disease by sniffing dead or dying badgers, or infected faeces or urine patches in the fields. The separation of cattle and badgers now seems to be the main hope of control the disease. This can be achieved by fencing off badger sets, avoiding hard grazing in areas occupied by badgers to prevent cattle having to graze close to their excreta, preventing badgers from feeding from cattle food stores and, in extreme cases, culling badgers. The maintenance of a high level of biosecurity on cattle farms should now be a major priority in high-risk areas, such as the south-west of Britain.

The new threat to the cattle population comes at a time when the public is much more sensitive to issues such as the use of snares or gassing to cull badgers. Hence, the emphasis is currently on improved husbandry and the development of a vaccine, rather than badger culls. In addition, cattle in susceptible herds need to be tested annually and reactors eliminated. Currently they are tested at frequencies of 1–4 years, depending on the severity of the disease in the area. The sensitivity of the tuberculin test is about 90% and its specificity even greater. It is unlikely that an increased frequency of testing would reduce the prevalence of the disease, but a new blood test, using gamma interferon, will help to detect the 10% of cases that might escape detection by the tuberculin test.

Cattle-to-cattle transmission is quite rare in the British Isles, accounting for only an estimated 15% of herd breakdowns. This is mainly because the test and slaughter policy prevents the disease reaching the fulminating stage where cattle become highly infectious. However, it does occur following cattle movement and in contiguous herds, although this may be because there are common territories of an intermediate host. Other countries have had considerable success, most notably Australia, by adopting a rigorous test and slaughter policy, although the success of this policy is reduced by the presence of a widespread intermediate host. In cattle, the incubation period is usually about 6 months, although some excretion has been noted in the early stages of infection. There is an increased risk of transmission from purchased bulls (Griffin and Dolan, 1995), which may relate to the stress that they suffer during transport (Knowles, 1999). There is evidence that this and other mycobacterial diseases of cattle, such as Johne's disease (Hutchinson, 1988), are more likely to emerge when housing conditions are poor and cattle are stressed. Contact with faeces is probably an important means of transmission and is frequent in modern scraped floor housing systems and following slurry spreading in fields. The spread of such diseases may ultimately require farmers to find ways of separating cattle from their faeces, such as slatted floors or regular automatic scraping of passageways.

Finally, as with many cattle diseases, there is a genetic component to susceptibility. Certain family lines are particularly susceptible and a breeding programme could reduce the susceptibility of cattle and help to control the disease in a manner that the public finds more acceptable than eliminating the supposed intermediary host (Phillips *et al.*, 2000).

Trypanosomiasis

Trypanosomiasis is a disease caused by the trypanosome parasite that is transmitted by the tsetse fly, which inhabits most of central Africa. The parasite causes intermittent fever, listlessness, progressive emaciation and eventually death. The distribution of the tsetse fly controls the livestock distribution in Africa, with only wild game and trypanotolerant cattle breeds inhabiting the heavily infected areas, such as in wet, swampy regions. In the open savannah, the tsetse flies prefer the wild game, but increasingly these have been replaced by cattle, as these are more valuable for food production. The desirability of controlling cattle trypanosomiasis has led to an extensive search for means of controlling the tsetse fly.

The disease is a particular constraint to the productivity of recently imported exotic cattle in Africa. Over several thousands of years, breeds of local cattle, such as the N'Dama and West African Shorthorn, evolved their own resistance, but these are not as productive as modern European cattle. Wildlife are carriers and do not suffer severe clinical symptoms, but they do provide a constant reservoir of disease organisms, rendering eradication of the disease impossible. The potential exists to transfer the resistance of the local cattle to more productive European cattle. If quantitative trait loci for trypanotolerance can be identified, it should be possible to transfer the relevant regions of the genome and produce novel genotypes with favourable disease resistance and production characteristics. The major challenge, however, is to understand the physiological basis for trypanotolerance, because it is only this understanding that can reduce the virulence of the disease in the long term. Clearly, the trypanosome haemoprotozoans are capable of commensal relationships in some cattle and wild animal genotypes, and this should be the objective of current breeding programmes for more productive cattle. Reliance on trypanocidal drugs and vector control has diminishing effectiveness, which is prompting considerable interest in breeding disease-resistant stock.

Herd Health Assurance Schemes

There is increasing concern among the general public that farm conditions are not always satisfactory for the health and welfare of cattle. The other major concerns that the public may have are for food quality and environmental quality. Some of these concerned members of the public are prepared to pay more for cattle products that have been produced to an assured health standard. Standards may be set by animal welfare charities, veterinary associations or large traders, such as the major supermarket chains, and by the government for specific diseases, e.g. enzootic bovine leukosis. Membership of such schemes may be required by the major supermarkets. Such standards will usually focus on the health of the herd and individuals within it, but also on related issues, such as hygiene on the farm, the quality of housing, plant and equipment, feedstuff and water storage facilities, stockmanship and the ability of a farm to manage an emergency. Monitoring is on a regular basis and may be done by veterinarians or more usually by staff specifically trained for the task, who have a checklist to examine different parts of the farm to assess their adequacy. The underlying principles of most herd health assurance schemes are to ensure that the cattle have:

1. Adequate space, in particular to allow for sufficient exercise indoors.
2. Freedom from aggression by other cattle, e.g. by providing adequate feeding and drinking facilities
3. Adequate floors to walk on.
4. A comfortable and clean bedded area.
5. Regular veterinary care.
6. Competent supervision by stockpeople.
7. Adequate transportation away from the farm where necessary.

Herd health assurance schemes are useful not only to monitor disease frequency, to enable comparisons to be made with acceptable standards, but also to determine risk factors contributing to disease and to implement control measures to improve performance.

Notifiable Cattle Diseases and the Laws Concerning Cattle Health and Disease in the UK

When diseases are of occasional occurrence, centralized action may be justified to contain the spread of the disease and prevent it becoming established as endemic in the population or even developing into an epidemic. Usually such action is taken by government authorities and in

Britain, the notification of 11 major cattle diseases to the Ministry of Agriculture is mandatory (Table 6.2), so that action can be taken. This may include compulsory slaughter, isolation of the site where the disease was found and action to contain the disease within the vicinity.

The spread of some cattle diseases is controlled by law in most countries. In Britain, there are 21 regulations that relate directly to the health and disease status of cattle (Table 6.3). The most important are the Agriculture (Miscellaneous Provisions) Act 1968, under which animals may not be subjected to unnecessary pain or suffering on agricultural land. The Act is supported by codes that describe the requirements for farm animals to avoid pain or suffering. Infringement of the codes is not in itself an offence but may be used as evidence that an animal has experienced pain or suffering. A separate code is available for cattle and all the other major farm-animal species.

Operations on farm animals are controlled by the Welfare of Livestock (Prohibited Operations) Regulations Act 1982. By this regulation, hot branding and tail docking of cattle are prohibited, as well as penis operations in the male and tongue amputations in calves.

The Welfare of Animals (Slaughter or Killing) Regulations 1995 requires that slaughterhouse workers should be trained in slaughter technique and should be licensed. They should also have a knowledge of animal behaviour and handling.

The Protection of Animals Acts 1911–1988 protect animals from danger and cruelty. For example, farmers can be banned from keeping cattle for periods from 6 months upwards. The Protection of Animals (Anaesthetics) Act limits operations without anaesthetics to injections or extractions (e.g. of blood), castration of bull calves up to 2 months of age, life-saving or pain relieving actions, experiments authorized by the Scientific Procedures Act 1986, and minor operations carried out by a veterinary surgeon quickly and painlessly (not including dehorning, castration or disbudding of calves).

The Welfare of Livestock Regulations 1994 (amended 1998) contains specific requirements for maintaining the welfare of livestock. Calves may not be kept in single pens unless certain pen-size minimum standards are

Table 6.2.   Notifiable cattle diseases in the UK.

Anthrax
Foot and mouth
Rabies
Warble fly
Tuberculosis
Pleuropneumonia
Enzootic bovine leukosis
Rinderpest
Brucellosis
Bovine spongiform encephalopathy
Blue tongue

Table 6.3.   Regulations controlling the health and disease status of cattle in the UK.

The Agriculture (Miscellaneous Provisions) Act 1968
The Animal Health Act 1981
The Animal Health and Welfare Act 1984
Anthrax Order 1938
Bovine Spongiform Encephalopathy No 2 Order 1988
Diseases of Animals Acts
Feeding Stuffs Regulations 1988
Markets (Protection of Animals) Order 1965
Medicines Act 1968
Milk and Dairy Regulations 1959
Protection of Animals (Anaesthetics) Act 1954, 1964 and Amendment Order 1982
Protection of Animals Acts 1911–1988
Scientific Procedures Act 1986
Transit of Animals (Rail and Road) Order 1975
Tuberculosis Order 1973
Votorinary Surgeons Act 1966
Warble Fly Order 1989
Welfare of Animals at Slaughter Act 1995
Welfare of Animals during Transit Act 1992
Welfare of Livestock (Cattle and Poultry) 1974
Welfare of Livestock (Prohibited Operations) Regulations 1982
Welfare of Animals (Slaughter or Killing) Regulations 1995
Welfare of Livestock Regulations 1994 (amended 1998)
Zoonoses Act 1975

conformed to. The diet must contain adequate minerals and fibre. There are also minimum space requirements for older calves housed in groups. This illustrates the more specific nature of recent legislation, as a result of research to identify which husbandry practices most severely affect welfare.

At the end of the 20th century, the desire for further improvement in the disease status of cattle and limitation of the spread of disease, particularly zoonoses, led to the introduction of cattle-tracing systems, which were mandatory in the EU member states after 1999. Under the system operating in Great Britain, cattle are required to have 'passports', which contain details of each animal's breed and sex; its date of birth and, eventually, its death; its dam's number and any movements that the animal has made throughout its life, and any government financial support that has been received for the animal. The scheme enables government authorities to trace cattle easily if there is a disease outbreak and to assure members of the public that the authorities have control of cattle movements. The system is managed centrally by the British Cattle Movement Service (BCMS), and paid for by the industry through a passport fee. The passport must include details of all of an animal's movements and BCMS must be notified either electronically or by post within 7 days each time an animal moves. When the animal is slaughtered, the passport is returned by the abattoir to the BCMS.

Conclusions

Never before have cattle diseases received so much attention from the public, veterinarians and farmers as in the UK during the last 20 years. The farming industry has been repeatedly criticized for failing to provide a healthy product for consumption, and the government has been accused of not safeguarding human health and of not acting on scientific advice. From the difficulties that have been encountered, there is emerging a system of cattle monitoring and health care that should ensure a significant reduction in the risk to consumers, providing that the very considerable economic damage to the industry does not limit the farmers' ability to implement the new measures. The systems put in place, such as the movement scheme and farm assurance schemes, should serve as a model to other countries wishing to reassure their consumers that beef and dairy products are safe foods to eat.

References

Blowey, R. and Edmondson, P. (1995) *Mastitis Control in Dairy Herds*. Farming Press, Ipswich.

Bramley, A.J. (1992) Mastitis and machine milking. In: Bramley, A.J. (ed.) *Machine Milking and Lactation*. Insight Books, Reading, pp. 343–372.

Byford, R.L., Craig, M.E. and Crosby, B.L. (1992) A review of ectoparasites and their effect on cattle production. *Journal of Animal Science* 70, 597–602.

Donnelly, C.A., MaWhinney, S. and Anderson, R.M. (1999) A review of the BSE epidemic in British cattle. *Ecosystem Health* 5, 164–173.

Griffin, J.M. and Dolan, L.A. (1995) The role of cattle-to-cattle transmission of *Mycobacterium bovis* in the epidemiology of tuberculosis in cattle in the Republic of Ireland – a review. *Irish Veterinary Journal* 48, 228–234.

Hemsworth, P.H., Barnett, J.L., Beveridge, L. and Matthews, L.R. (1995) The welfare of extensively managed dairy cattle – a review. *Applied Animal Behaviour Science* 42, 161–182.

Hutchinson, L.J. (1988) Review of estimated economic impact and control of Johne's disease in cattle. *Agri-Practice* 9, 7–8.

Knowles, T.G. (1999) A review of the road transport of cattle. *Veterinary Record* 144, 197–201.

Krebs, J.R., Anderson, R., Clutton-Brock, T., Morrison, I., Young, D. and Donnelly, C. (1997) *Bovine Tuberculosis in Cattle and Badgers*. MAFF Publications, London.

Laevens, H., Deluyker, H. and de Kruuf, A. (1998) Somatic cell count (SCC) measurements: a diagnostic tool to detect mastitis. In: Wensing, Th. (ed.) *Production Diseases in Farm Animals*, Proceedings of the 10th International Conference, Utrecht. Wageningen Pers., Den Haag, pp. 301–310.

Owens, F.N., Secrist, D.S., Hill, W.J. and Gill, D.R. (1998) Acidosis in cattle: a review. *Journal of Animal Science* 76, 275–286.

Phillips, C.J.C. (1997) Review article: animal welfare considerations in future breeding programmes for farm livestock. *Animal Breeding Abstracts* 65, 645–654.

Phillips, C.J.C., Foster, C., Morris, P. and Teverson, R. (2000) The role of cattle husbandry in the development of a sustainable policy to control *M. bovis* infection in cattle. Report to the Ministry of Agriculture, Fisheries and Food. MAFF, London.

Politiek, R.D., Distl, O., Fjeldaas, T., Heeres, J., McDaniel, B.T., Nielsen, E., Peterse, D. J., Reurink, A. and Strandberg, P. (1986) Importance of claw quality in cattle – review and recommendations to achieve genetic-improvement. Report of the European Association of Animal Production Working Group on Claw Quality in Cattle. *Livestock Production Science* 15, 133–152.

Further Reading

Blowey, R.W. (1990) *A Veterinary Book for Dairy Farmers*. Farming Press, Ipswich.

Blowey, R.W. (1993) *Cattle Lameness and Hoof Care*. Farming Press, Ipswich.

Brand, A., Noordhuizen, J.P.T.M. and Schukken, Y.H. (1996) *Herd Health and Production Management in Dairy Practice*. Wageningen Pers., Wageningen.

Stanford, C.F. (1991) *Health for the Farmer*. Farming Press, Ipswich.

University Federation for Animal Welfare (1999) *Management and Welfare of Farm Animals*, 4th edn. UFAW, Wheathampstead.

West, G. (1992) *Black's Veterinary Dictionary*, 18th edn. A&C Black, London.

Housing, Handling and the Environment for Cattle

7

Introduction

The housing of cattle is an economic necessity in many parts of the world, though not to the same extent as the housing of poultry and pigs. Dairy cows are most likely to be kept indoors for a large part of the year, because of greater control of diet, the need to limit damage to a farm's pastureland, and the opportunity to mechanize milking and other aspects of the routine animal care.

The overall aim in housing cattle is to provide an economic system of production with high labour efficiency. This aim emphasizes that, whatever is desirable for the wellbeing of stockpeople and animals under their charge, the housing system must be financially viable. Some people might anthropo-morphically believe that cattle are always more contented in a more natural environment outside. In some situations, this is undoubtedly true. However, for a high-yielding cow grazing sparse pasture in winter, or a cow kept outside her thermoneutral zone, this may not be the case. Studies where cattle have been given the choice of indoor or outdoor environments have demonstrated that cattle will choose to remain indoors during inclement weather or when feed availability is greater than outside. Remembering that we have extensively modified *our* environment to improve our comfort and the facilities available to us, it is wrong to imagine that cattle are always better kept outdoors.

The specific objectives of cattle housing are:

- to provide a comfortable environment and adequate food and water supplies for the animals, thus meeting their behavioural and physiological needs;
- to provide a comfortable and safe working environment for the stockperson;
- to minimize injury to stock and the transfer of diseases;
- to provide ready access for the cows to handling facilities and in the case of lactating cows, the milking parlour;

- to protect the land area of the farm from damage by cattle treading or overgrazing.

Much planning should go into the design of cattle housing, because it is an infrequent investment and it can be costly to rectify mistakes. Cattle houses must be designed with future requirements in mind.

- How profitable will the cattle enterprise be in relation to other enterprises?
- Is it desirable to change the size of the herd, or the breed?
- Will the breeding policy increase the size of the cattle?
- Should there be other facilities for the cattle e.g. for dairy cows, should the milking parlour expand?

The positioning of future expansion should be considered, so that an efficient and profitable unit is eventually arrived at. Future building developments must be anticipated. Recent materials advances have produced lightweight, large span buildings without any central supporting pillars. These allow much greater flexibility, with more space for large machinery to be used inside the building, and better ventilation.

The designer of cattle buildings also has to try to anticipate the changes in input/output cost structure and the legal requirements for cattle farming. The following points should be considered.

- Will the proposed unit be economical in its use of resources that are increasingly valuable, such as water?
- Can excreta be efficiently moved away from the cows and treated?
- What are the desired levels of mechanization and labour to service the building?
- Are noxious odours released close to human habitation?
- Is the welfare of the cows constrained by building design?

The Cattle House

The most important elements of any cattle house are the floor, the lying area and the feeding system.

Floors in cattle buildings

The floor is the physical point of contact of the animal with its environment, and is important from the point of view of wear and tear on the animal, primarily the hooves, the ability to sustain normal locomotory behaviour and the conduction of heat from the animal to the floor. Floors must be designed to withstand the heavy animal traffic caused by a high stocking density in the house. The intricate balance between a grass sward and the soil would soon be destroyed by the high stocking density in cattle houses. To withstand the traffic, floors have to be hard, much harder than most outdoor surfaces, and they should also be non-absorbent. Concrete is the material of choice, since it is relatively durable, inexpensive and not too slippery, at least in the initial period after laying. It can be laid with a variety of types of corrugated surfaces that help to prevent the animals slipping, from a tamped surface, which is created at the time of laying by stippling the surface with a plank of wood, to a grooved surface, which is usually created with a cutting device in floors that have been worn smooth over time.

Slipping is a particular problem for lactating cows on smooth floors, and the cows may not be able to get up if their legs splay. Inflatable bags positioned under the cow may assist her to rise to her feet, but the problem is better addressed by taking preventive measures. Falls, slips and splays on smooth floors cause bone, muscle and joint problems, but a concrete floor that is too rough can lead to damage to the sole of the hoof. A high risk of slipping reduces a cow's welfare and mounting activity at the time of oestrus. Slipping is particularly a problem on concrete because the hard surface prevents the hoof from sinking into the floor surface as it does on earth. On concrete, the heel bulb deforms, increasing the surface area in contact with the ground and the force applied to the hind part of the hoof. Eventually the heel bulb may be eroded, causing the cow to walk on the hind part of the hoof with the toes losing contact with the ground and becoming overgrown. Frequent scraping of concrete floors to remove slurry soon leads to the ridges created by tamping being worn away.

Slipping can be minimized by increasing the coefficient of friction of the floor, determined as the force required to move an object over a floor divided by the weight of that object. Cows have few problems on floors with a coefficient of friction above 0.4, but with less friction, there is a rapid increase in slip frequency (Fig. 7.1). The risk of slipping is greatest at the beginning of the stride (just before the thrusting phase), when the forward horizontal force of the cow is large relative to the vertical force of the cow's mass, and the friction provided by contact between the hoof and the floor is reduced.

Particular attention should be paid to floor quality in areas of heavy cattle traffic, such as around water troughs and in the feeding area. In dairy cows, heavy traffic occurs in the milking parlour, and in places where cows are required to turn sharp corners suddenly, such as in leaving a building to go to be milked or entering or leaving the parlour. As cows turn a corner, the outer and inner limbs rotate, putting more pressure on the outer and inner claws of those respective limbs. In high-risk areas, the floor can be treated with an aggregate embedded into a resin, which reduces slipping considerably.

Other environmental and cattle factors will influence the likelihood of cattle slipping:

• wet surfaces are more slippery than dry surfaces, so regular removal of surface water is preferable;
• freshly tamped surfaces are rather better than grooved surfaces at providing a slip-resistant surface;

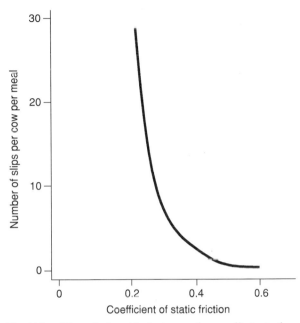

Fig. 7.1. The relationship between the coefficient of static friction and the number of slip movements of four cows measured during four eating periods.

- cows with small upright feet are more likely to slip than cows with large, overgrown toes, which means that care must be taken in breeding cattle with 'improved' hoof conformation.

Slatted floors

These are made from a number of parallel concrete beams (slats) with gaps or slots in between. Poorly constructed slatted floors cause cattle to walk with their head down, fixing their gaze on the floor ahead of them. This enables them to position their hooves carefully. There is less walking activity in total and it is slower than if cattle are on solid floors. Cattle may even spend less time grooming their hindquarters because of the risk of over-balancing. However, well-designed slatted floors allow cattle to be kept at high stocking densities without bedding and to remain reasonably clean.

If slats are too narrow, there is an unacceptable strain on them, particularly if they are very long, but if they are too wide, there is inadequate disposal of faeces between the slats. Optimum slat width is about 150 mm and the slats should be T-shaped to encourage dung to fall into the pit below (Fig. 7.2). The gaps between the slats should be about 40 mm wide. If they are less than this the faeces does not pass through easily. Great care should be taken that slats have been manufactured from high-grade concrete to the

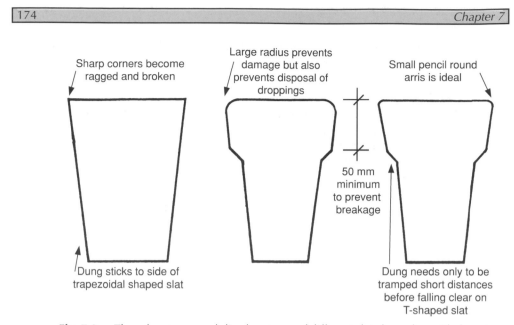

Fig. 7.2. The advantages and disadvantages of different slat shape for cattle housing floors (courtesy of General Concrete Products Ltd, Newcastle-upon-Tyne).

required loadings for the class of stock to be kept on them, including the necessary reinforcement.

Lying areas and feeding

Cattle, like humans, spend about one-third of their life lying down and resting. Lying is important to cows; it allows them time to recuperate and digest their food (Metz, 1985). Digestion is a long process because their forage contains much structural carbohydrate, and bacterial fermentation is required before the nutrients are in a suitable form for absorption.

Cattle do not spend more than about 5 min at a time sleeping, which can be recognized by their neck being recumbent, with the head usually tucked back so that it rests against their thorax (Ruckebusch, 1972). In the wild, the lack of sleep probably maintained awareness and reduced the chance of attack by predators. Up to about 8 h a day are spent ruminating, in a semi-trance-like state, which may substitute for sleep.

In most countries cows are kept inside and fed conserved feeds for at least part of the year, because grass and other crops for grazing will only grow in warm wet conditions at certain times of year. Inside, cows can be accommodated in areas where they are free to move around and lie down (loose housing), or they can be tethered by a chain or with their head between two bars in individual stalls (tie stalls). If loose-housed, cows usually have access either to food delivered along a passage (easy feeding) or they help themselves from a clamp of silage or from racks of feed outside (self-feeding; Ingvartsen and Andersen, 1993).

Feeding along a passage

If cattle get their food offered to them on a concrete passage, they must be prevented from walking on it by a barrier. The design of this barrier is important, as the cows will strain to get feed that is furthest from them and which has not been sampled by other cows. As a result, they exert a lot of forward pressure on the barrier. It is also important that the design of the barrier prevents the cows taking food and retreating with it to their lying area, since there will be some waste of the food as it falls from their mouth to the floor. To achieve this, the barrier may be in the form of tombstone-shaped units, usually made of wood, which force the cow to raise her head high before stepping backwards (Figs 7.3 and 7.4). Alternatively, sloping metal bars can have the same effect, with cows having to turn their head sideways to move away from the barrier (see also Fig. 7.4). The base of the barrier should be firmly secured to the ground, otherwise the force of cows feeding will move it forwards.

The food for the cattle should be provided either on the passage floor or in a trough. Unless at least 0.2 m per adult cow of feed trough is provided, there will be much aggression at feeding time. Cattle prefer fresh forage and not that which has been recently contaminated by other cattle, a defence against disease transmission. Hence, when the food is put out for the cattle, they fight to get first access.

Cattle often take a mouthful of feed and throw it forwards into the middle of the passage (food tossing), particularly if the food is not presented at floor level but is in a raised trough (or bunk in the USA). It is a time-consuming job to fork it all back to the cows, but an angled scraper can be mounted onto a

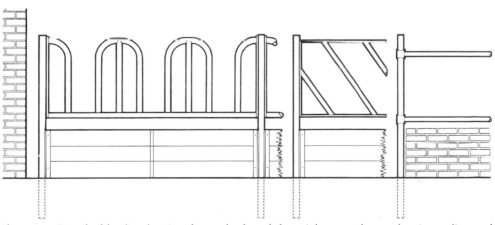

Fig. 7.3. Detail of feeding barriers for cattle, from left to right: a tombstone barrier, a diagonal barrier and a simple twin-bar barrier.

Fig. 7.4. Photograph of feeding barriers in use by cattle, from left to right: a tombstone barrier, twin-bar barrier, diagonal barrier and sloping barrier.

tractor to return the feed to the cows. Food tossing also makes the cows' backs dirty, and wastes food. If the food is in a trough at floor level, cattle find it harder to toss their food forwards, but it will rarely hold enough food for more than one-half to 1 day. A horizontal wire or bar on an open barrier will stop them food tossing, but will need to be raised up for young cattle as they grow. The aetiology of the disorder is believed to be because the animals do not spend a long time tearing grass from a sward, as they would have to do when grazing (taking 30,000–40,000 bites of the sward per day). It may also be related to selection, as they avoid food that has been sampled by other cows, which they can easily sense with their well-developed sense of smell.

Self-feeding
Cattle take their food, usually silage, directly from a store, which is usually clamped between two walls. It is essentially a process of vertical grazing. A barrier is needed in front of the clamped silage to prevent wastage, so that the cattle cannot trample on feed that has fallen to the floor after being taken from the clamp. The herdsperson should aim to keep an even face of silage, which will be easier if the material that has been ensiled is all of similar quality. Cows readily refuse silage that is mouldy, contaminated, overheated or has undergone a fermentation producing butyric acid, for example.

   The barrier for self-feeding silage clamps is either free-standing, in which case it is likely to comprise a metal pole attached to a wooden frame, or it may be an electrified metal pole that is suspended from a bar inserted into the silage. This bar has to be hammered further into the silage as food is removed from the clamp, preferably about 15 cm day^{-1} to avoid the silage undergoing a secondary fermentation when it is exposed to air. The secondary fermentation is temperature-dependent. A roof cover for a clamp of silage is not essential but it does keep the rain off the silage and allows the area to be used for other purposes if required. Many covered clamps have been converted to cattle housing as herds expand. One disadvantage of self-feed silage is that silage can only be put into the clamp up to a height of about 2 m, because this is the maximum height that cows can reach to remove the silage (depending on their size), whereas if silage is removed mechanically this can easily be increased to 3 m, providing that the clamp walls are high enough and the guide rail is visible from the tractor during loading. A third possibility is to make a tall clamp and remove the top layer with a block cutter to be fed to the cows in a circular feeder, of the type used to feed round bales of silage or straw to cows.

Both covered and uncovered outdoor clamps that are used for self-feeding should be lit at night for the cows to see what they are eating, otherwise intake will decline. It is particularly important with outdoor clamps to light up the route that cows should take to the clamp. Lights can be easily controlled by photoelectric sensors, some of which can be set to come on at different light intensities. In high rainfall areas, the cattle often get wet and dirty if they feed outside, increasing the time required to clean them in the parlour. The silage in an uncovered clamp can also get very wet if the cows are not eating through the silage rapidly, and this could reduce the intake of dry matter.

Usually food in clamps is compacted with a tractor and then covered with plastic sheet to keep out the air that would allow aerobic bacteria to spoil the silage. If the silage in the clamp is finely chopped, the compactness may make it difficult for some cows, particularly young animals that are losing their milk teeth, to remove the food from the clamp. At the feed face, there may be quite a lot of bullying of younger cows, which may be reluctant to feed if there is an electric barrier anyway.

The cubicle

In cubicle housing, cows are given access to raised lying beds of c. 2 × 1 m, usually with an absorbent material (bedding) on the surface. They can walk into and back out of these beds, but they should not be able to turn around on them. The beds are separated from each other by a division, usually constructed with metal bars. At best, a cubicle division creates a barrier between neighbouring cows and increases the feeling of personal space. At worst, it acts as a restriction to movement of the cow, especially when it lies down and gets up, and may lead to damage to the cows' legs. Some farmers believe that a solid division between the ends of cubicles placed end-to-end is preferred by cows, since this restricts their view of cows opposite them. If no barrier is present, direct respiratory contact may increase the risk of transmitting respiratory disease, such as tuberculosis.

The cubicle division is usually made of metal or wooden bars to give the cows more freedom of movement and to reduce cost. If the cubicle is too big, the cow may try to turn around rather than back out to leave the cubicle.

This is particularly true for young animals not used to lying in cubicles. They dislike backing out of cubicles into a busy passageway, as this exposes their vulnerable flanks and udder to attack by other cows, and on rising may be tempted to turn around. Similarly, cows may get stuck in the front of a cubicle if they shuffle forward, since they need to be able to lunge forward as they get up (Fig. 7.5). Sometimes a 'brisket board' is positioned on the floor at the front of the cubicle to stop cows going too far into the cubicle and becoming stuck. If there is a big variation in cattle size within a herd, some young heifers may get stuck in oversized cubicles as they turn around, but large old cows may find the same cubicles too small. This is one reason for not bringing young heifers into the herd when they are very small. If cattle have difficulty getting up, they may reverse the normal pattern of behaviour, choosing to get up with their front legs first instead of their back legs. They may even sit 'doggy-fashion' in the cubicle for some time, which is a sign of unsuitable cubicle construction.

The division between cubicles is critical to the successful provision of lying space for cows. It usually has one or two points of insertion into the cubicle base. Metal bars often corrode at this point, so it is advisable to paint them or use a plastic sleeve to protect this part of the cubicle. The division can have a bottom rail to stop cows invading their neighbours' space. The height of this is critical – if too low, cows get their legs trapped underneath; if too high (which is more likely), they may roll underneath and injure themselves when attempting to get free. Standard heights are about 350–450 mm for the lower rail and about 1000 mm for the top rail. Some cubicle divisions have the lower rail replaced with a twisted rope. This will stretch under pressure from cows if necessary, but it still prevents the cow from straying into her neighbours' space too much. The Dutch Comfort cubicle division minimizes the length of the lower rail (Fig. 7.6), and allows some space-sharing between

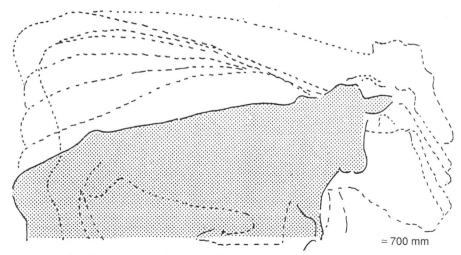

≃ 700 mm

Fig. 7.5.   The forward space demand of rising movement for an 800-kg Friesian cow.

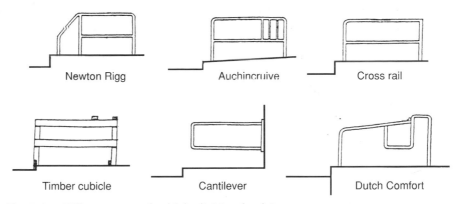

Newton Rigg   Auchincruive   Cross rail

Timber cubicle   Cantilever   Dutch Comfort

Fig. 7.6. Different types of cubicle division for dairy cows.

cows (Phillips, 1993). A U-shaped tubular steel section is positioned close to the wall to prevent cows moving into the next cubicle. The position of this section from the wall should be no more than 350 mm, otherwise young heifers can turn into the hole and get stuck. Cows in Dutch Comfort cubicles lie more laterally recumbent, whereas those in more confined cubicles are sternally recumbent. In the field, sternal recumbence is the usual means of repose for adult cows, but there they can change position frequently, which is difficult in cubicles.

Attached to the cubicle division, but at right angles to it and 450–500 mm from the front of the cubicle, a neck rail is usually secured. This forces the cow to back out of the cubicle as she stands up, and if she defecates as she does this the faeces will fall onto the floor and not onto the cubicle base. If positioned too far from the wall, it is uncomfortable for cows when getting up.

The cubicle base should be slightly sloping (a 70–80 mm fall over the length of the cubicle) to allow urine to drain off into the passageway and to allow cows to lie uphill, which relieves pressure on the diaphragm on a hard surface. The base is usually made of 100 mm-thick concrete on top of a well-consolidated base, providing a bed that is at least 200 mm high, to prevent slurry being pushed onto the bed during scraping. If higher than about 250 mm, cows are reluctant to enter and it is uncomfortable when they stand with their front legs on the cubicle and back legs in the passage, as well as putting more weight onto the rear hooves, which may increase lameness. Solid concrete bases should be insulated with tiles or polystyrene beneath the surface in cold climates. The base can also be made of bitumen or rammed earth. Cows find the latter comfortable, as they hollow out a bed so that it is moulded to their shape better than flat concrete, but fresh earth must be added at regular intervals as it gets removed by the action of the cows' feet into the passageway. Recent attempts to improve the comfort of cubicle bases for cows include, firstly, waterbeds placed on top of the base and, secondly, putting old car tyres into a hollowed out base and covering it with an impermeable material.

The cubicle base should be covered with a bedding material. In addition to improving the cow's comfort as she lies in the cubicle, the bedding should also cushion the impact of the cow as she lies down, absorb moisture from urine and provide a clean surface for the cow to lie on. The base may have a lip or kerb 50–75 mm high at the end, to stop bedding being removed from the cubicle, but if the cubicles are too short, it is uncomfortable for cows to lie on the lip and they may prefer to lie in passageways. The use of small straw bale choppers to chop the straw to 5–10 cm has reduced this problem and prevented slurry pumps becoming choked with long straw. Straw is the most common and usually the cheapest form of bedding, and should be provided to a depth of 50 mm. In practice, the centre of the cubicle quickly becomes devoid of straw as it gets pushed to the side by the movement of the cow, so fresh straw should be provided at least three times a week and preferably daily. It is not very absorbent, but it is free of bacteria. Moist sawdust can harbour mastitogenic organisms, such as *Klebsiella* species, and both straw and sawdust can create a dust hazard when they are added to the cubicles. Where available, baled wood shavings are quite acceptable but usually more expensive than straw. Shredded paper or newsprint has also been used, but it can cause ink contamination of the surface of the cows' udders. Sand can be used to a depth of 50–100 mm, but it requires a lip on the base to assist retention and is likely to cause wear in the slurry pump. It is not absorbent. Increasingly, farmers are installing mats and carpets to provide a permanent solution to cubicle bedding. Mats are comfortable but expensive, and they may need regular cleaning at the joins between the cubicles. Carpets have the advantage that they can be rolled out under the cubicle divisions, avoiding the need for joins, but they quickly get compressed and need additional bedding material for adequate cow comfort. Many farmers resort to using bedding and mats. Soiled litter at the rear end of the cubicle should always be removed at least two or three times a week, and the herdsperson should regularly sprinkle a small amount of lime there to sterilize the area where the udder may be challenged by mastitogenic bacteria.

PASSAGEWAYS   Cubicle passageways should be at least 2.2 m wide to allow cows to pass comfortably behind others that are standing half out of the cubicles. The feed passage should be wider, at least 2.8 m, to ensure that cows can pass freely behind other cows that are feeding. Slurry should be removed by a tractor-mounted rubber scraper at least once a day and preferably twice, if possible during milking when cows are out of the building. The less disturbance to the cows the better. Automatic scrapers that are attached to a heavy-duty chain keep the passageways very clean, but they wear the concrete more rapidly to leave a slippery surface. They are not an acceptable way of making cows lie in cubicles. Care should be taken that cows tails are not trapped in the scraper as it passes down the passageway, by having a trip-out device installed into the electric motor. Areas that cannot be reached by tractor-mounted or automatic scrapers should be cleaned by a hand-operated scraper daily.

Passageways should be ideally be arranged so that there are no blind alleyways where subordinate cows can be trapped by dominant cows. Frequent cross passages between the cubicle and feed passageways should be provided to encourage a good flow of cows in the building. Relative to other environments, there is a lot of aggression between cows in a cubicle house. Nearly always, this is highly ritualized, with the dominant cow swinging her head in the direction of the subordinate cow, who moves out of the way, or at least lowers her head to indicate that she accepts the dominant status of the other cow (Phillips, 1993). The more-extreme forms of agonistic behaviour are mainly seen at pasture, where there is little ritualized aggression. At pasture you will see cows engage in head to head contact, wheeling around in a test of strength, with the victor eventually gaining access to the vulnerable flank areas of the vanquished cow. Such an overt display would be dangerous in a cubicle passageway, which is usually slippery and full of other cows. The thwarting of aggressive interactions may induce a certain amount of tension, leading to cows seeking hiding places, such as in the cross passages or half-in and half-out of a cubicle. This may induce a sense of security for a cow because awareness is centred in her head region.

Strawed yards

Strawed yards provide cows with free access to an area with deep, soft bedding, but there are no individual beds for the cattle. They are often used for growing male beef cattle, as cubicles are unsuitable for males, who would urinate in the middle of the cubicle bed, and also for growing cattle, for whom a constant cubicle size would be impractical. Strawed yards for beef cattle are frequently uncovered, it being difficult to justify the cost of a roof on financial grounds alone. When the straw is mixed with faeces, urine and large amounts of rainwater, it turns into a sludge, and if allowed to accumulate over the winter, the cattle may sink up to their hocks as they attempt to walk through it. It is difficult to speculate on the effect of such systems on the welfare of the cattle – freedom of movement is restricted and the matting of hair when it is covered with muck is probably uncomfortable, it being a high priority for most animals to keep themselves clean. In the UK, traditional farms often had a portion of the yard covered, which gave the cattle somewhere dry to lie in wet weather. Many of these units are difficult to manage as the straw cannot be removed by machine, and they are gradually being phased out of service, despite the durability of the buildings constructed for this purpose in the 19th and early 20th century.

Many strawed yards, particularly those for dairy cows, include a concrete passage next to the feeding trough. This provides a clean hard surface for them to stand on while feeding, and is useful to allow cows in oestrus access to an area where no cows are lying down. It also makes it easier to get lactating dairy cows out of the building for milking. If there is no concrete passage, the high frequency of treading and frequent defecation and urination make the straw wet and the cows dirty. A concrete passage also helps to provide an abrasive surface to wear away the hoof growth that proceeds unchecked over the winter in strawed yards (see Chapter 6 for a comparison of leg disorders in strawed yards and cubicles.)

Increasingly, strawed yards have been advocated for dairy cows to improve their welfare, which is a testament to the failure of many cubicle systems to provide adequate comfort. When given the choice, nearly all cows prefer strawed yards. The two main advantages to the cow of being accommodated in a strawed yard rather than a cubicle house are that the bedding is deeper (and therefore she is more comfortable), and she has greater freedom of movement. Cows often spend several hours longer each day lying down in strawed yards than in cubicles. The greater freedom of movement brings its own dangers, in that cows in oestrus may accidentally tread on lying cows, and particularly on exposed teats, sometimes tearing the teat wall. For a lactating cow, this is painful, and a high stocking rate should be avoided to minimize the problem. Oestrus behaviour is more exuberant in strawed yards than in a cubicle house and is easier for the herdsperson to detect (Phillips and Schofield, 1990). Mounting activity is more likely to be prolonged and there is more pelvic thrusting by the mounting cow. Lordosis (a curved spine that indicates the standing reflex) is particularly evident in the cow being mounted. This is probably because of the availability of 'safe' space on the hard-standing area, with better footing and freedom of movement. Cubicle houses were not designed with the oestrus cow in mind: the floors are usually slippery and there is a danger of cows interacting with the cubicles on dismounting. The close proximity of the cows in a cubicle house encourages sexual activity, compared with cows at pasture, for example, but they may be frustrated by the environment and reluctant to mount each other.

Regular provision of adequate straw is vital to good management of a strawed yard. Usually at least 1 t per cow is required over a British winter, and some farmers use 2–3 t per cow if it is readily available from an arable enterprise. Providing more straw keeps the cows cleaner, thereby reducing the time required in the parlour to clean the cows and the risk of environmental mastitis. This amount of straw is about 5–10 times the amount used in a cubicle system. Over a period of several weeks, the straw heats up as micro-organisms grow and release heat. If the straw is not removed, the heat will eventually sterilize the composted mixture in the lower regions of the straw bed. Some farmers leave the removal of straw to the period when the cattle have just been turned out at pasture, which is often one of the quietest times of the year, but in this case, the strawed yard must be designed to accommodate a rise of up to 1 m in the surface level of the straw. This may involve raising any dividing gates or barriers. It is, however, better for the health of the dairy cows if the farmer removes the straw/excreta mix on a more regular basis, say every 3 weeks. In this way, the bed can be kept relatively clean and mastitis is less likely. The environmental implications of disposing of animal excreta as slurry are discussed in Chapter 9.

CUBICLES AND STRAWED YARDS – SPACE ALLOWANCES   Cows in strawed yards are usually provided with 8–9 m² each and those in cubicle housing about 6 m² each. By contrast, grazing cows have several thousand square metres each, and beyond 360 m² seem to form a stable spatial relationship to each other. Grazing cows in a large group normally keep about 10–12 m from their nearest neighbour, whether they are grazing or lying down. When lying, cows in cubicles are usually less than 1 m from their neighbouring cow, who may or may not be their preferred partners, and those in strawed yards are usually within about 2 m of the nearest cow. Clearly, we expect housed cows to tolerate a much closer presence of other cows than in the field.

Any determination of the minimum space requirements for cows in strawed yards should take into account the cows' need for lying space, ventilation and space for walking and performing oestrus behaviours. The space required for walking and oestrus may be provided by a hard-standing area, which is often a concrete passage where the cows stand to feed, but may also be provided by an outside dirt-lot area. An allowance should be made for accumulation of the straw in the lying area when constructing the division between the concrete feeding passage and the lying area, by creating a step onto the lying area which is of sufficient height. The following space allowances should be the minimum provided (Table 7.1), except where the cows are kept in high ambient temperatures, such as in Israel where the bedded areas are often doubled to allow adequate ventilation around each cow. The yard should preferably be rectangular, with the feeding trough down the long side. If the trough area per cow is small, i.e. less than 0.2 m per cow, it will increase the treading around the access to the feeding passage, if one exists, making the straw dirty and leading to an increase in mastitis.

One cubicle should be available for each cow, unless the cows are in a large group, in which case there may be a limited opportunity for the number to be reduced on the assumption that not all cows will want to lie down at once. Most cows have favourite cubicles that they prefer to occupy and for this reason it is better if there is at least one cubicle available for each cow. Broken divisions should be promptly mended, so that the number of available cubicles is not reduced for any period of time.

Table 7.1. Space allowances for cows in strawed yards.

Live weight (kg)	Bedded area (m² per cow)	Hard-standing (m² per cow)	Total area (m² per cow)
600	5.50	2.00	7.5
650	6.00	2.00	8.0
700	6.25	2.25	8.5
750	6.50	2.50	9.0

The size of the cubicle bed should relate to the average weight of the cows, although in herds where the heifers come into the herd at a low proportion of their mature weight it may be that the largest cows will find it difficult to get into a cubicle designed for an 'average' cow, and at the same time the smallest cows can turn around and perhaps get themselves stuck. However, assuming the herd is quite uniform in size, the dimension of the base should be determined from the weight of the average cow in the herd (length in metres = 1.75 + 0.00068 × weight in kg; width is one-half of the length). Adequate length is most important and existing cubicles that are too short can sometimes be lengthened by putting wooden sleepers at the end, providing that it does not make the passageway too narrow.

TIE STALLS   In most parts of the world, cows were traditionally tethered in stalls, and this system is still common in some large herds in Eastern Europe and many small herds in traditional production systems. With the need to reduce labour use to make the dairy industry more efficient, as well as the changes in animal husbandry systems in Eastern Europe after the political changes of the 1990s, tethering in cowsheds is becoming less common. Some farmers have had to abandon tethering because of the high labour requirements for feeding cows individually. Cows in tie stalls also tend to have poor reproductive performance, compared with loose-housed cows.

In stalls, cows are either tied by the neck with a chain or kept with their head in a yoke, the former giving the cow more freedom of movement. They can get up and lie down but not turn around. An electrified wire (cow trainer) is sometimes suspended just above the cow's back to encourage her to move backwards as she arches her back to defecate or urinate, so that excreta falls in the passage behind the cow's bed. The cow trainer reduces contamination of the bed and hence mastitis, but can also restrict the cow's movement. In the UK, tie stalls were usually made shorter than cubicles for loose-housed cows, only about 1.5 m, to ensure that excreta fell into the passage and not onto the back of the cubicle bed.

Usually there is a simple partition between the cows, often of solid wood to reduce draughts. The cows are fed in troughs at the front of the stall, either made of concrete or tiles so that they can be easily cleaned, or more tradition- ally wood, which is difficult to keep clean. Water is usually provided from a small bowl in the stall, which is triggered by a noseplate. The cows are milked in their stalls, usually nowadays with the milk passing directly to a pipeline, and the milking unit transferred between cows. Sometimes the milking units are suspended from a gantry, which reduces the labour requirement. Otherwise milk is collected into cans, and then transferred to churns or a bulk tank for collection.

Inevitably, tethering cows restricts their freedom of movement, which infringes most modern welfare codes. However, there is little or no aggression between cows and it should not necessarily be assumed that cows prefer to be loose housed. In many traditional cowshed systems, the bond between the stockperson and the cows in his charge was much stronger than it is in loose housing systems, which may have compensated to some extent for the limited contact between animals. The frequent contact between loose-housed cows, if of an aggressive nature, may reduce their welfare. However, the restriction of movement of cows in tie stalls can cause leg disorders, particularly swollen knees and hocks if there is insufficient bedding or the stalls are too short, and will usually cause joint stiffness. The UK's Farm Animal Welfare Council recommend that cows are untied and allowed to exercise daily.

Loafing areas

In some countries, an outdoor exercise area, without straw, is provided with housing for either dairy cows or beef cattle. This loafing area or 'dirt lot' reduces the stocking rate and gives cows in oestrus somewhere to mount each other safely. In hot climates, the loafing area may be combined with a shade providing roof.

Dry cow and calving accommodation

Dry cows are best housed in a strawed yard, at least those within 2–3 weeks of calving, as this gives them the freedom of movement that they need during late pregnancy. In addition, the projected calving date may not be accurately known and a cubicle house is far too restricted an environment for cows to calve in. The cows should be regularly inspected for signs of impending calving. Just before calving, they should be moved to an isolation box, which may also be used for sick cows. These should be about 4–5 m^2, giving the cow some room for manoeuvre during parturition, and there should be one box for every 10–20 cows, depending on the spread of calving. In summer, it is better to allow the cow to give birth at pasture, preferably isolating her in a paddock of her own. The incidence of postnatal disease is usually less when the cow calves outside, emphasizing the importance of keeping the calving area clean and disinfecting between calvings. The floor and walls of the calving area should be of impervious material for easy cleaning and there should be a deep bed of straw provided. It should be well lit and free from draughts that might chill the newborn calf.

Out wintering

Fewer cows in Britain are now left outside all winter than 50 years ago, because of the improved techniques for conserving forage as silage to feed them in winter and better housing systems. Also cows are now stocked at high rates on pasture because of increased grass growth as a result of improved varieties and more fertilizer use, and this increases the risk of poaching damage to the pasture in winter. Nevertheless, for a long time farmers have recognized the benefits of letting cows out to pasture for a short time each day in winter, often between morning and afternoon milking. This gives the cows some exercise and makes a useful contribution to their diet, as well as reducing the amount of excreta to be stored and disposed of safely. In Britain and countries with similar climates, it would be unwise to leave pasture ungrazed in autumn in order that it be available for winter grazing because of the risk of frost damage. However, many winters support continued grass growth that is best removed if a good spring flush of growth is to be obtained. The aim should be to provide only young, leafy grass at all times for the dairy cow.

Permanent housing of dairy cows

In most countries, cows are kept inside and fed conserved feeds for at least part of the year, because grass and other crops for grazing will only grow in warm wet conditions at certain times of year. Permanent housing is increasingly favoured because of the greater control of the diet that is achieved, and a widespread adoption of the new automatic milking systems would further reduce the number of farms offering their cows access to pasture. However, permanent housing is prohibited in organic dairy systems, so it is important to recognize the possible problems that can ensue.

Cattle behaviour in permanent housing

FEEDING BEHAVIOUR   At pasture cattle normally graze for 8–12 h day^{-1} and take 30,000–40,000 bites from the pasture each day. Depriving cows of the opportunity to perform this behaviour by feeding them conserved feed, which can be consumed in about one-half of the time, can lead to abnormal behaviours. One of these is feed tossing, described previously. Other abnormal oral behaviours demonstrated more by housed cows than by cows at pasture include tongue rolling and mammary sucking.

SOCIAL BEHAVIOUR   There is an increased frequency of aggressive interactions in housed cows compared with grazing cows, but these tend to be ritualized and incomplete. The reduced distance between neighbours indoors makes all forms of social interactions more likely, including grooming (which may help to pacify animals close in the dominance order), oestrus behaviour and fighting. The cubicle division probably acts to reduce this social contact, increasing the cow's perception of personal space, but the low level of comfort provided by most cubicles results in the cows spending less time lying down than in strawed yards.

ACTIVITY LEVELS   Cows walk at least twice as far every day when they are kept at pasture than when they are housed in cubicles or a strawed yard. Some of this walking is necessary for food selection purposes and hence apparently redundant if food does not have to be harvested by the cow itself, but there may be a need for a certain level of activity each day, which is thwarted when cows are kept permanently indoors.

Permanent housing and cattle health

MASTITIS   Repeated contact of dairy cows with dirty bedding leads to increased mastitis. There is usually an increase in infection by *Corynebacterium pyogenes*, often in association with other pathogenic bacteria such as *Peptococcus indolicus,* during wet, warm periods (summer mastitis), with *H. irritans* (the headfly) as the vector. In summer, the proliferation of organisms that cause environmental mastitis, particularly coliform organisms, is increased in cattle housing because of higher temperatures. It is good practice to clean and disinfect dairy housing when the cows are turned out to pasture, and the period of rest, together with the sterilizing effect of sunlight during the summer, will produce a clean environment for the cows to be housed in the autumn. There may not be an increase in mastitis with permanent housing if the cows have access to an exercise paddock. Note that 'summer' mastitis commonly occurs in permanently housed cows in the second part of the winter, suggesting that an accumulation of bacteria may predispose cows to the disease.

LAMENESS   The repeated contact of the hoof with wet acid slurry predisposes cows to heel necrosis (underrun heel) and digital dermatitis. In heel necrosis, the acidity of the slurry and the proteolytic enzymes erode the heel bulb, producing a pitted area which may penetrate the sensitive tissue and lead to infection. It also predisposes the cow to laminitic disorders. The constant wetting of the heel predisposes to the condition since the hoof moisture content is increased and the tissue softened.

Some hoof disorders, most notably white line separation and punctured soles, are the result of small stones becoming embedded in the sole, usually during passage down a stony track to and from milking. On farms with stony, and especially flinty, soils these conditions can be avoided by permanent housing, although other options, such as adding a new surface to the farm tracks, are available to overcome the problem.

With year-round housing, there is an increased incidence of laminitis and to a lesser extent solar ulcers, but this may be partly caused by the high level of concentrates that must be fed to compensate for the lower nutritional value of silage compared with fresh grass. Lameness also depends on physical conditions and social factors. In deep-strawed buildings, hoof wear is less than in cubicle buildings, a problem which is often overcome during the summer period when cattle walk further than when they are indoors. If cattle are permanently housed in strawed yards, it is likely that lameness will be reduced, but regular trimming will be necessary to control excessive hoof growth. Care is necessary to control bacterial accumulation in the straw, as interdigital dermatitis is more common. Often this is initiated by interdigital hyperplasia caused by straw damaging tissue in the interdigital cleft

NUTRITION   All-year housing usually requires the feeding of conserved feeds, since in most areas, zero-grazing is not possible for part of the year, such as periods when the soil is too wet for harvesting machinery to operate or the land is too dry or cold for cattle food to grow. The conservation process has been improved with the transition from hay to silage, the development of silage additives to expedite anaerobic conditions and the introduction of rapid harvesting and ensiling procedures to ensure minimum respiration and ensiling losses. Nevertheless, the intake of silage by cattle is still less than the fresh material from which it was made, which is probably because of the presence of protein breakdown products in the silage. Consequently, supplementary concentrates are needed to avoid low milk production and possible reproductive failure. This can increase the hoof problems referred to previously, particularly laminitis, as a result of endotoxin release. During ruminal acidosis, the release of histamine is also stimulated and both the endotoxins and histamine can damage blood vessels, which in the hoof restrict blood flow and the supply of nutrients to the hoof corium. Poor quality hoof tissue results, which is susceptible to injury and infection.

Another potential problem of year-round housing is a low vitamin E status of the cows, which may predispose them to mastitis.

REPRODUCTION   In some cows that are housed all year, there are unacceptably low reproduction rates, because of, in part, low condition scores. Oestrus is potentially easier to observe indoors and the close proximity of cows encourages a more active oestrus display, provided the floor surface provides a good grip and the cows have adequate space.

Ventilating Cattle Sheds

Even in the most extreme cold climates where cattle are kept, such as in Canada, the heat produced by the fermentation of feeds digested in the rumen is sufficient to ensure that the temperature inside the cattle house is sufficient without artificial heating. Artificial ventilation is required in some cattle buildings to regulate air temperature, and to control relative humidity and noxious gases at high stocking densities. However, for adult cattle ventilation is normally by the natural influx of external air, which rises as it is warmed and leaves the building by the ridge. This is known as the stack effect, which ventilates the building by the vertical movement of airstreams of different temperatures as a result of convection (Fig. 7.7). Air that enters the building falls, as it is colder than internal air. As it is warmed and rises, it takes with it pollutants that have accumulated around the animals. In winter, the internal/external temperature difference is greater and air falls faster and further, creating a risk of chill to young animals near the point of air entry. Air usually enters the building under the eaves or through Yorkshire (slatted) boarding on the upper half of the walls. It leaves through an open ridge or slots cut in the roof. The ridges of cattle roofs are often not sufficiently open to allow

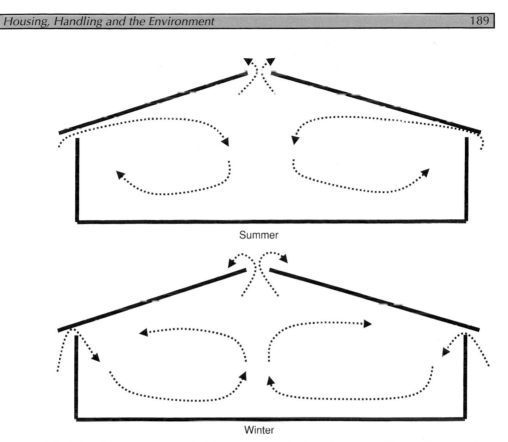

Fig. 7.7 The movement of air in summer and winter in a naturally ventilated cattle building.

adequate air exit and some air may leave through the slatted boarding of the walls. In very large single-span buildings, there may not be sufficient upward lift of the waste air to allow it to reach the ridge, so a number of smaller double-span buildings are preferable for adequate air movement.

If artificial ventilation is required, in calf buildings in cold climates for example, then it should be remembered that it is difficult to overcome the effects of high stocking rates by artificial ventilation. For instance, the benefit in reduced microbial population of doubling the air space per calf (i.e. halving the stocking rate) can only be achieved by a fivefold increase in the air change rate (Webster, 1984).

Lighting

Daylength

Most people take artificial light for granted, since we enjoy a well-lit environment for most of the day regardless of the natural daylength. By contrast, many cattle are kept in houses without artificial light or with only the minimum provision to make inspection of the animals possible in the dark. Sometimes this is deliberate, as with veal calves, for whom a low light intensity discourages excessive activity, thereby enabling them to cope with their environment better.

At extreme latitudes the daylength in winter may be less than 8 h. Cattle prefer to perform some activities in the light, in particular feeding since food selection may be difficult without a clear sight of what is on offer. If supplementary light is provided, cattle utilize the extra time to spread out their feeding activity over the day, but the total amount of feeding time is not affected. The same occurs with grazing cows as daylength increases in mid-summer – cattle spread out their meals over the daylight hours. For dairy cows, walking and feeding in a crowded cubicle house may be stressful in the dark, since the ritual aggressive displays that normally maintain the dominance hierarchy and the distance between cows become more difficult. There is evidence that providing supplementary light alleviates some of the stress of living at a high stocking density in a cubicle house. The supplementary lighting should cover the feeding area as well as the lying area, otherwise cows are encouraged to stay in the cubicles and do not go as often to feed. The transition from a lit area to an unlit area can be uncomfortable for a cow and it probably takes their eyes about as long as ours to adapt to the dark, about 15–30 min.

Daylength perception is relative in cattle and indeed most mammals, and in dairy cows a declining daylength, i.e. in autumn, appears to be more stressful than one which is short but not declining. The optimum daylength to produce the highest milk yields in dairy cows is 16–18 h, allowing the cow a quiescent period of 6–8 h at night. This may also be optimum for the cow's welfare. Lighting affects the cows' lying and in particular sleeping behaviour. Cattle sleep less in winter if they are given supplementary light. In natural light, they normally sleep for longer in winter than other seasons, which before domestication would have had adaptive advantage in conserving energy and reducing the risk of predation.

Daylength also affects the cow's reproduction. Cows calving in spring usually have delayed conception compared with cows calving at other seasons of the year, because, if they did conceive in spring, they would give birth in mid-winter when, during the course of cattle evolution, less food was available. However, this effect is not nearly as strong as, for example, it is in sheep, which are seasonally anoestrus. Providing supplementary light in winter advances puberty in heifers, as it does in several other species. It also reduces the rate of lipogenesis, but not growth rate, so beef cattle slaughtered in winter will have leaner carcasses if supplementary light is provided. Cattle use declining daylength as a cue to store fat to help them survive the winter, and for this purpose, nutrients are diverted from muscle and bone growth. In lactating cows, the fat content of milk is also reduced by supplementary light provided in winter, and we can surmise that pre-domestication, it was advantageous for cows in the short daylengths of winter to produce milk with a high fat content for their calves.

Intensity and colour

At present, the optimum intensity for cattle productive or welfare purposes is not known. The physiological effects (such as increased prolactin concentrations and decreased melatonin) are likely to be triggered by intensities as low as 1 or 2 lux. However, higher intensities encourage increased activity levels, since they provide cattle with a more visually comfortable environment. Perhaps more important than the exact intensity is the means of providing the supplementary light, as some high-intensity spot sources of lights, e.g. halogen lights, could reduce the welfare of cattle by the glare they produce. Using a small number of high-intensity lights could also produce large contrasts in illumination intensity across a building, especially if they are not mounted high enough above the cattle living area. The alternative of installing a large number of fluorescent tubes is expensive, and sodium lights may be a better option. These are more efficient but their light output is more orange than normal daylight. Different colours almost certainly cause differences in the animal's mood and this could be used to improve their environment. Although early indications are that cattle, like other mammals and some birds, are more active in red than blue or green light, the long term effects of keeping cattle in different colours are not known. Red is commonly found to incite arousal in other species and this is a universal phenomenon that is used to good effect by bullfighters. In the early stages of a Spanish bullfight, the matador uses a large fuchsia-coloured cape, which probably stimulates the bull with its movement. In the final stages, he changes to a small red cape held directly in front of him. This focuses the bull's attention for a final, well-directed charge, whereupon the matador attempts to plunge a sword into the bull's spinal cord just behind the neck.

Cattle are able to distinguish red colours from either blue or green but so far, no-one has been able to demonstrate that they can distinguish blue from green. The ability to distinguish red from other colours may relate to the detection of the reddening of the vulva during oestrus. It could also be useful in enabling cattle to assess the quality of grass from its colour, since it normally changes from green to yellow as it ages.

Perception of visual stimuli

The above considerations indicate that there are many unknown aspects of the effects of light stimuli on cattle. Some indication of the likely effects can be provided from a knowledge of their perceptive faculties, even if we are currently unable to examine the extent of possible processing. Of major relevance to visual perception is the positioning of the eyes on the side of the head, giving cattle about 330° vision, compared to our 180°. This was obviously an advantage for prey animals, and all stockpeople should know that approaching cattle from behind provides no guarantee of remaining unseen. One consequence of having a wide visual field is limited binocular overlap, about 40° in cattle, which may restrict the extent of their depth perspective at close quarters.

The region of best visual acuity in cattle is spread over a wide area, about 130°, in comparison with less than 1° in humans. This is achieved by having a 'visual streak', where retinal ganglion cells are concentrated in a horizontal line across the retina. This enables cattle to have good vision on the horizon. When combined with motion parallax (positioning of moving objects by the overlap of monocular images), it gives an excellent ability to detect moving objects on the horizon.

Cattle possess an effective tapetum, or reflective layer behind the retina, which effectively allows light to be counted twice as it passes through the retina. This suggests that they have good vision at low light intensities, probably better than our own. Nevertheless cattle are still reluctant to graze at night, which is probably a vestigial defence mechanism, and they take longer selecting mouthfuls when they do graze at night.

Housing Pollutants

Excreta

Slurry creates an unwelcome environment for cattle, and they go to quite extreme lengths to avoid lying in, or walking through, deep pools of slurry. As well as producing noxious odours, chiefly ammonia and sulphur compounds, it is acidic and moistens the hoof, leading to a high rate of abrasion on concrete, particularly of the heel. Slurry is normally removed from the passageways either by being scraped out of the building once or twice a day, using a rubber strip mounted on the back of a tractor, or by an automatic scraper in each passageway attached to a heavy-duty chain that moves down the length of the passage every few hours. Alternatively, cattle may be kept on slatted floors, which allow the slurry to fall between the slats into a pit. In such systems, a high stocking rate is essential, otherwise dried faeces accumulates in the slots. Slurry in pits has to be emptied regularly and care should be taken of the gases, particularly hydrogen sulphide, that are emitted especially during mixing of the slurry.

The recent trend towards large areas of concrete in and around cattle accommodation may be short-lived because of the increased volume of dirty water created that has to be disposed of in a way that is sympathetic to the environment. One recent development that illustrates this problem is the rapid-exit milking parlour, which usually has a large apron of concrete around it, which the cows traverse as they leave the building. In future, it may be necessary to restrict the area where excreta are deposited to limit the area to be washed and dirty water accumulation.

Noise and vibration

Cattle hear high-frequency sounds much better than humans, their high-frequency hearing limit being 37 kHz, compared with only 18 kHz for humans (Heffner and Heffner, 1993). Their best audible sound is also at a higher frequency, at about 8 kHz, compared with 4 kHz for humans. Their low frequencies hearing limit is similar to that of humans (about 25–30 Hz). Despite being better at hearing high-frequency sound, cattle have less ability to pinpoint the direction from which it comes (localization). Humans can localize a sound in a horizontal sphere to within an arc of 1–2°, but cattle can only localize it to an arc of 30°. The practical inferences are that some high-frequency noise that we cannot hear will be audible to cattle and it may disturb them, but they will not be able to determine where it is coming from to take evasive action. However, thresholds for discomfort are unlikely to be breached in normal circumstances. These thresholds have been determined for other mammals, but not cattle, and are of the order of 90–100 dB, with physical damage to the ear occurring at 110 dB. Some noise may indeed be welcomed by cattle, providing interest in an otherwise dull day. Music in the milking parlour, for instance, has been shown to encourage cows to enter, but this could be because of the pacifying effect on the herdsperson, of whom they are then less frightened.

Vibration from machinery or heavy traffic may cause discomfort, particularly in cows lying on concrete floors that are close to a road with heavy traffic. Low frequency vibrations from large vehicles travel further than high frequency vibrations and are more likely to be a disturbance, in the same way that low frequency noise is less easily attenuated than high frequency.

Volatile compounds

The main volatile substances that can be harmful to cattle are noxious pathogens and odours, although in some circumstances pheromones from nearby animals can provide an unwelcome distraction. Problems in cattle houses are much less than pig or poultry buildings because of the lower stocking density. In calf houses, build-up of pathogens presents a particular problem and in cold regions, they may have to be ventilated artificially. Of the range of noxious odours that occur at high stocking densities, ammonia is the most common and is created by the volatilization of nitrogenous compounds in excreta. Like hydrogen sulphide, it causes irritation of the eyes and throat in concentrations in excess of 400 ppm. It can also irritate the mucous membrane of the respiratory tract, leading to reduced pulmonary clearance of bacteria (Wathes and Charles, 1994).

Microbes

A wide range of microorganisms, including bacteria, fungi and plasmids, can contaminate the aerial environment of cattle and are a particular risk factor in calf diseases. Some of the organisms are zoonotic and are therefore a cause for concern for humans as well. Poor air hygiene is a major contributory factor in the complex web of calf respiratory diseases. In temperate climates, it is best to ventilate calf houses naturally, while taking care to avoid draughts at calf level. Solid walls are, therefore, essential to at least a height of 1.5 m. The air inlet area should not be as great as in adult cattle houses since the calves do not generate sufficient heat for an effective stack effect. Too large an influx of cold air will fall rapidly on entry, particularly in winter when there is a large difference between ambient and internal temperatures.

Adequate disinfection is essential between batches of calves, and in a block calving herd this usually means after the main period of calving. Both calf pens and the building should be cleaned with a high-pressure hose, disinfected with an iodophor or chlorine-based detergent and rested until needed for the next batch of calves. If this is not done, there is likely to be an accumulation of contamination and a rapid spread of calf diseases. All-in–all-out systems of calf rearing, where pens are disinfected between batches, usually have fewer losses than continuous flow systems of calf rearing. This should be considered in designing new calf accommodation, and there is an increasing tendency to occupy calf accommodation continuously in some of the large dairy farms. Calf hutches are increasingly believed to provide a more suitable microclimate for the calf, but the restriction of movement and contact with other calves is contrary to high welfare standards.

Dust

Dust particles are disliked by cow and calf alike. Dusty food, either forages or concentrates, is universally avoided. The dust particles are created by sloughed skin particles, bedding and feed, and include both inhalable dust, that is deposited in the respiratory tract via the nose and mouth, and respirable dust that is deposited in the exchangeable region of the lungs. Both are potentially dangerous because of pathogens or mutagenic/allergenic substances that can be carried into the respiratory tract. Bovine epithelial and urinary antigens can invoke antibody responses in people working with cattle and are also associated with allergic reactions. Of particular importance is extrinsic allergic alveolitis, or Farmer's Lung, which is derived from dusty food and can affect both cattle and stockmen, but the transition of many dairy farms from making dry foods (predominantly hay) to wet ones (predominantly silage) has reduced the prevalence of this disease. Farmers, however, are known to have a particularly high incidence of respiratory disease, of which Farmer's Lung in one example. Organic dust is also dangerous because of the fire hazard it creates when it accumulates around light fittings.

Milking Facilities

For the dairy farmer, milking is a time-consuming task. Most cows are now milked mechanically, although hand milking prevails in some developing countries. Cows are usually walked to a specialized building for milk collection, the parlour, which must be situated close to a room for cooling the milk and storing it in a tank. Since the tank must be accessible from a road, this constrains the layout of dairy farm buildings.

The first attempts to mechanize the milking of cows were made in the 19th century, by inserting cannulae into the teat canals. Some applied pressure to the outside of the udder to stimulate milk let-down, and in the latter part of this century vacuum began to be applied around the teat.

Mode of action of milking machines

The expression of milk from the teat can be achieved by two means – squeezing and sucking. Squeezing is the main method used by the calf and during hand milking. Pressure is applied to the base of the teat by the calf's tongue pressing the teat against its upper palette (this is alternatively done by the milker's fingers during hand milking) and this pressure is passed down the teat, causing evacuation of the teat cistern. Milking machines, however, rely on evacuating a closed area around the teat at regular intervals (about once a second). The calf also applies some vacuum by enclosing the teat in its mouth. The maintenance of the correct pulsation rate (number of cycles per minute) and pulsation ratio (ratio of vacuum level to atmospheric level) are important in minimizing teat damage and optimizing milking efficiency.

Closure of the teat canal at the end of a cycle is achieved by the pressure exerted by the collapsing liner. It is important for teat condition that the glands are not overmilked, and that the full vacuum is achieved for at least 15% of the cycle, and that a sufficiently low pressure is achieved, i.e. not above 50 kPa (Mein, 1992).

Milking machine components

The main components of any milking machine are the milking cluster, which comprises four teat cups that apply the vacuum around the teats; the vacuum system, which includes the vacuum pump and line, which connects to the teat cups; the pulsator, which alternates the applied vacuum with atmospheric pressure to prevent the teat being damaged; and the transport pipeline to take the milk to be cooled and stored (Fig. 7.8).

The vacuum system is protected by an interceptor jar, which prevents liquid entering the pump; a sanitary trap (in pipeline systems), which prevents contamination of the vacuum system with milk; a regulator, which maintains a steady vacuum, and finally a vacuum gauge, which allows the efficient running of the vacuum to be monitored.

The milking cluster, four teat cups with soft liners (usually of synthetic material), is connected by short milk tubes to a clawpiece that collects the milk from the four teat cups. The clawpiece admits air to break up the milk column for easier transfer, and acts as a weight to keep the four teat cups in the correct position. Milk is transferred from the four teat cups by a long, flexible milk

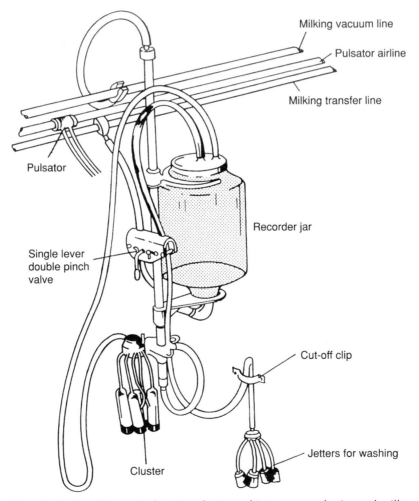

Fig. 7.8. Typical milking unit showing cluster and jetters, recorder jar and milk, and vacuum transfer lines (from Whipp (1992), courtesy of Insight Books, Ltd).

tube to a fixed pipeline for transfer to a bulk tank (direct-to-pipeline). It may pass via a recorder jar, which allows the yield at each milking to be recorded, or flow meters, which are inserted into the long milk tube with the same purpose.

Milking parlours

Tie stalls

Cows that are still kept permanently tethered in stalls (tie stalls) are usually milked by milking units, which are moved between them. The milk is either collected into a can suspended under the cow's body or into a churn that is wheeled down the cowshed passageway. In some of the more sophisticated systems in use in large units in Eastern Europe, the milk is taken from the cow by a milking machine that travels down a gantry running the length of the passageway. As in a parlour, milk passes to a pipeline that conveys it under vacuum to a bulk tank. More labour is required to milk tethered cows than those that walk to a parlour.

Conventional milking parlours

Some of the first milking units to be introduced did not rely on moving the cows to a parlour for milking but taking the milking system to the cows at pasture. However, as more farms were established with a central farmsteading and good access to the farm's grazing by cowtracks, static milking parlours were developed in the steading. Initially 6–8 cows were arranged side-by-side in stalls (the abreast parlour, often on a raised platform, and they left the milking place by the front of the stalls, thus allowing the next cow to enter from behind. However, milking a large number of cows in this way is tedious because of the bending down that is necessary to access the cow's udder. Therefore, the tandem parlour was developed (Fig. 7.9), where cows enter and leave the stalls either side of a pit, which enabled the herdsperson to stand upright. In modern tandem parlours, exit and entry gates can be automatically opened after the cluster has been removed, allowing the cow to leave and the next one to enter. This allows cows to sort out their own order of entry into the stalls, leading to cows being more contented during the milking process.

The major drawback of large tandem parlours for modern dairy herds is that the herdsperson must walk long distances if there are many stalls. Thus, a tandem parlour does not normally have more than three or four stalls either side of the pit, which limits the use of this type of parlour to herds of 100 cows or less. The greater the number of milking units the faster the milking time for the herd, provided the milker can use the extra units effectively.

Before the development of automatic entry and exit gates, a modification of the tandem parlour, the chute parlour (Fig. 7.10), was developed which allowed cows to enter and exit each side of the parlour as a single group. This saved building space. This design was further developed to allow cows to stand at an angle of 30–35° to the side wall, reducing the distance between

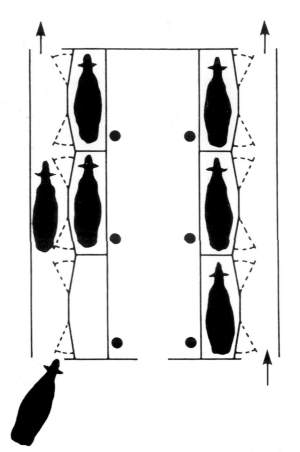

Fig. 7.9.   The tandem parlour (from Whipp (1992), courtesy of Insight Books, Ltd).
•=milking unit.

cows' udders and therefore the walking by the milker. This popular design, called a herringbone parlour, could be extended with additional milkers in the pit, and some herringbone parlours now have 48 units in two rows of 24. Initially, milking units with a central pit were shared between the cows on either side of the pit and were passed from one side to the other after a cow had finished being milked. Although this allowed for efficient use of the milking units, milkers were often standing idle waiting for cows to finish being milked. Most parlours now have a milking unit for each cow place. The limiting factor for milking speed is not availability of a milking unit but cows that are slow to release their milk. Large parlour units with long lines of cows suffer most.

Recently, trigon and polygon parlours have been developed for large herds, where blocks of four to six cows stand in lines at a 35° angle to the wall in a triangular (trigon) or diamond (polygon) configuration (Fig. 7.11). The system allows each line to have its own entry and exit passage. Such units

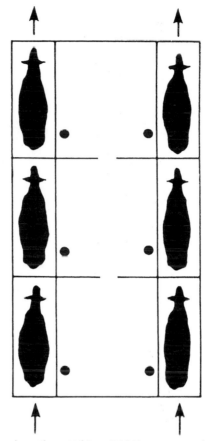

Fig. 7.10. The chute parlour (from Whipp (1992), courtesy of Insight Books, Ltd).
•=milking unit.

provide large pits for a comfortable working environment, but they cannot be extended if a herd increases in size. An alternative to cows leaving by the end of the passage where they are standing to be milked is for the side wall or barrier to lift up, the 'rapid exit parlour'. The disadvantage of such a parlour is the large exit floor area that has to be cleaned, but it can avoid cows having to be hurried down the milking passage at the end of milking to allow the next row to enter.

In large farms, an alternative to static milking parlours is for cows to walk onto a raised turntable, which transports them past the milker, who therefore has little walking to do. These rotary parlours also offer the potential to have a small number of automation units, such as concentrate feed delivery units, which can be activated when the turntable passes a certain point. Some rotary parlours suffer from a high risk of breakdown, but most recently, they have been designed so that the turntable floats on water, thereby reducing the mechanical requirements and increasing reliability. Some cows

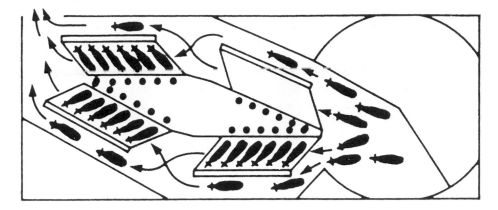

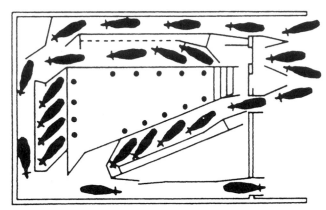

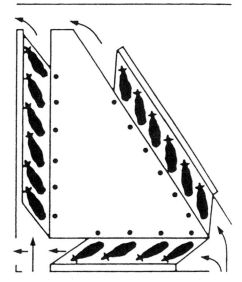

Fig. 7.11. Two examples of a trigon parlour (middle, bottom) and a polygon parlour (top) (from Whipp (1992), courtesy of Insight Books, Ltd). •=milking unit.

are reluctant to enter and leave a moving turntable, but various configurations can be used to ensure that the cow is moving mainly forwards (cows are difficult to train to back out of, or move sideways from, any space, let alone stepping from a moving platform to a static floor). In some units, the clusters are attached from inside the turntable – the tandem and herringbone rotary parlours – whereas in the rotary abreast, they are attached from the outside. This releases the milker so that he or she can encourage cows that are reluctant to enter the parlour, whereas the rotary tandem and herringbone may require two people, one inside and one outside the turntable. The advantage of access to the cow entry point in the rotary abreast parlour may be offset by the disadvantage of cows disappearing from view after the initial stages of milking.

The development of new milking parlour designs was accompanied by the introduction of automated operations concerned with milking. The following were the most significant.

CLUSTER REMOVAL   Sensors detect when the milk flow rate falls below about 0.25 kg min^{-1}. This activates a piston that pulls a nylon cord attached to the cluster and removes it from the cow.

YIELD RECORDING   Initially, the yield of cows was recorded by collecting the milk in a jar, which could be read by an operator working alongside the milker. Later, developments focused on the recording of either a proportion of total milk entering a meter and scaling it up to give total yield, or measuring the force generated to change the direction of flow of the milk as it passed – a continuous flow meter. Meters recording a proportion of the total yield may measure either the weight or volume of the sample taken. Despite the large variety of yield recording methods, they are usually accurate to within 2% of total yield, 0.25 kg if the yield is less than 10 kg.

MASTITIS DETECTORS   Cows with mastitis produce milk with an elevated sodium content which increases the conductivity of the milk. This can be detected with a pair of electrodes placed in the cluster or short milk tube, and the resulting conductivity measurement correlates closely with the somatic cell count. Handheld devices are also available. Milkers can get advance warning of an incidence of mastitis using this method, and antibiotic treatment will be more effective if given before the infection is properly established. However, mild cases often recover spontaneously, and care must be exercised when deciding whether to treat with antibiotics. Mild cases could be isolated and milked last, to stop the infection spreading.

Mastitis can also be detected by filters placed in a transparent container that is inserted into the long milk tube. Clots collect on the filter and the milker can then be alerted not to allow the milk to pass into the collection tank.

COW IDENTIFICATION UNITS   The first units to be developed were unsatisfactory because the reception range of the interrogating units was too short (3–5 cm), leading to many occasions when cows entering the parlour were not identified. New systems have a much greater range, and cows can be individually identified from the radiosignal emitted at unique wavelengths as they enter the parlour through an archway that generates an electromagnetic field. Once the cow has been accurately identified, a feeder can be activated to deliver a pre-programmed amount of feed or information can be downloaded on how far she has walked since the last milking (if the identification unit is on her leg).

ACTIVITY MONITORS   Activity monitors, or pedometers, record the number of steps taken by a cow by means of piezo-electric member or mercury switch connected to a real time microchip and individual identification unit. This can be attached to the cow's leg or suspended around her neck. Because the increase in activity during oestrus is proportionately greater and more consistent than changes in other parameters, such as milk temperatures, yield and vaginal mucus conductivity, the pedometer offers the best potential to automate oestrus detection. Theoretically, 100% accuracy can be achieved, with no false positives, but errors may be made if cows change their physical environment, e.g. from housing to pasture, or become ill, especially lame. Activity increases by a factor of about 350% on the day of oestrus, and an algorithm can be used to distinguish this increase from other, non-oestrus, variation (Phillips, 1990). The information may be relayed to a control processor, at milking time, along with cow identification, or it can be used to signal directly to the herdsperson that the cow is in oestrus, using flashing lights.

CONCENTRATE FEEDERS   Cows that are fed concentrates in the parlour are easier to collect from the field, but they are more excited during the milking process. The widespread adoption of out-of-parlour feeders in the 1980s, and more recently complete diets, has discouraged farmers from feeding their cows in the parlour. This can create difficulties when the cows are at pasture and a farmer wants to feed concentrates to individual, high-yielding cows.

Conventional parlour feeders rely upon the milker entering the cows' identification numbers as they enter the parlour, and a pre-programmed computer instructs the release of individual concentrate rations to feed troughs in each milking stall. The feed delivery devices have to be regularly calibrated to ensure accurate allocations to each cow. The maximum that adult cows can eat is $0.4 \, \text{kg min}^{-1}$, and rather less than this for heifers. This limits the total daily intake to 8–10 kg, which is insufficient for some high-yielding cows.

TEAT CLEANING AND DISINFECTION   Teat cleaning can be helped by the provision of a spray line in the parlour. In hot dry countries, cleaning can be more effectively achieved by automatic spray units set into the floor of the collecting yard. These also remove the dust from the udder and cool the cow before milking.

Teat disinfection can be automated by fitting a floor mounted spray line in the floor of the exit passage, triggered by cows interrupting a beam of light.

Chlorine-based disinfectants can be used in this case and more effective teat coverage ensured than if the milker quickly passes a spray line under each cow in the milking parlour.

AUTOMATIC CLUSTER ATTACHMENT (ACA)   ACA was the main obstacle to fully automated milking through much of the 20th century. Towards the end, sensors based on ultrasound, infrared or laser waves began to be used to locate the teats, mainly by experimenters in Holland, France, Britain and Germany. Some systems include a computer memory of the position of the teats in each cow. Following location of the teats, a robotic arm attaches the cluster from the side of the cow or it emerges from a false floor. The arm remains under the cow during milking and removes the teat cups when milk flow has declined.

The cow must be restrained in an individual, fully automated stall since ACA can only be used if the other milking operations are automatic. Cows can then be allowed to enter the milking stall voluntarily rather than under the supervision of the milker. High-yielding cows will visit three or four times each day, which will increase milk yield and hence milk production efficiency compared with a conventional twice-a-day supervised milking. A system must be in place to prevent the milk of cows with mastitis from entering the bulk tank. Cleaning the unit between cows may limit the spread of mastitis better than conventional parlours.

Fully automatic milking

The prospect of cows presenting themselves for milking voluntarily is now realistic, following a great deal of research which has been undertaken to develop a fully automatic milking machine. This is a very significant development in the automation of milking which will change the routine of cows and herdspeople. For the cow, it offers more frequent relief of udder pressure, reduced udder weight and less contact with the herdsperson. Being milked by a machine is an unnatural process and may be viewed as frightening for the cow if she is of nervous disposition, but better for the cow if the herdsperson instils fear in the animals under his or her control. A good herdsperson will help the cows to overcome their fear of the machinery, but rejection rates are likely to be greater for fully automated plants than conventionally 'manned' parlours. However, for the farm owner and herdsperson, the time saving with automatic milking is very attractive (Table 7.2) and may ultimately prove essential if the farm is to remain competitive (Kuipers and Rossing, 1996).

Automatic milking systems must be able to detect when a cow has mastitis and divert her milk away from the bulk tank. Cows should be in a position to visit two to three times per day, which may be difficult in some grazing systems. The cows may need to be encouraged to visit by providing concentrates at milking. More frequent milking will increase milk yield and the efficiency of milk production, as well as reducing the udder weight with potential benefits for hind limb locomotion. It is most likely to be

Table 7.2.   The effect of automation on cow milking routine time (min per cow) in the parlour.

	Basic routine	Automatic system
Let cow in	0.25	–
Foremilk	0.10	0.10
Wash and dry udder	0.2	0.2
Attach cluster	0.2	0.2
Remove cluster	0.1	–
Disinfect teats	0.1	–
Let cow out	0.2	–
Safety margin	0.05	0.05
Total	1.2	0.55
Cows per man hour	50	110

considered by large farms that can afford the high investment cost. However, the saving in labour should not be achieved at the expense of routine care of the cows.

Managing Cattle in Extreme Climates

Ruminant cattle produce considerable amounts of endogenous heat caused by the microbial digestion in their rumen, and they are therefore more prone to heat stress and less to cold stress than other farm animals. In common with other homeotherms, cattle increase evaporative heat loss (sweating, panting, etc.) as the ambient temperature increases and the sensible heat loss decreases (convection, conduction and radiation; Fig. 7.12 illustrates this for a calf). Below the lower critical temperature (LCT) and above the upper critical temperature (UCT) they invoke physiological mechanisms to maintain core body temperature (38.2°C), so heat production increases and the efficiency of milk production or growth is reduced. Milk production can be reduced by up to $5\,l\ day^{-1}$ in extreme high temperatures. Between the LCT and UCT is the zone of thermoneutrality or comfort zone, and in a controlled environment the most economic temperature for housed cattle is just above the LCT, where the artificial heat provision is at a point where there is minimum heat loss from the animals. Most cattle are, however, not kept in controlled environments, so that this would only apply to calves that are kept in environmentally controlled buildings in cooler regions.

The zone of thermoneutrality in adult cattle is from approximately −20 to +26°C, depending on the environment and animal factors that influence the critical temperatures. One of the most important of these for young cattle is the rate of air movement, as draughts remove the temperature shells surrounding a calf's body and rapidly chill it. For adult cattle that tend to be more prone to heat stress than youngstock, the productivity of the animal is most important, since a high-producing dairy cow produces a much greater

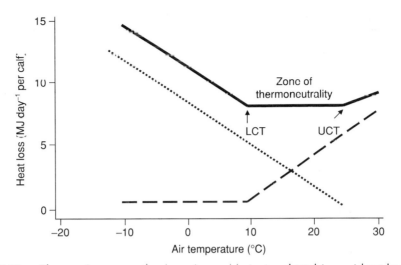

Fig. 7.12. Changes in evaporative (— —), sensible (····) and total (———) heat losses of a calf with air temperature. LCT=lower critical temperature; UCT=upper critical temperature.

heat increment because of increased digestion compared with a beef cow at maintenance.

In cold environments, the neonatal calf is particularly at risk of cold stress, even though it has reserves of brown adipose tissue (BAT) to produce heat (non-shivering thermogenesis). Compared with other farm livestock, the reserves of BAT are proportionately greater in calves than in lambs (piglets have no BAT at all, but are usually delivered into an environment where the temperature is close to that of the womb). Hence, neonatal mortality in the calf as a result of hypothermia is less of a problem than with lambs. The relatively large body mass of calves in comparison to lambs, and the lower surface area:volume ratio, helps calves to preserve body heat. BAT is laid down in the latter stages of pregnancy and rapidly metabolized after birth to help overcome the large temperature differential, perhaps 30°C, between the cow's womb and the environment. This gives the calf about 50 h to live if no food is consumed, compared with lambs and pigs, both of which have only 10–15 h. However, a calf must suckle well before 50 h if it is to absorb adequate immunoglobulins from the colostrum, preferably within 6 h.

The ease with which a calf suckles depends on both the mother and calf. The vigour of one or both may have been challenged by a prolonged calving, and in extreme cases of hypoxia, the calf may be severely weakened and unable to stand to suckle. In such cases, the drop in rectal temperature can last for 10 h or more (Fig. 7.13). One of the mother's first tasks is to stimulate the calf to suckle by massaging it and orientating it to her udder. This is achieved quite often by licking the calf's anus, which helps to expel the waste products that have accumulated in the calf's gut during pregnancy. Primiparous cows

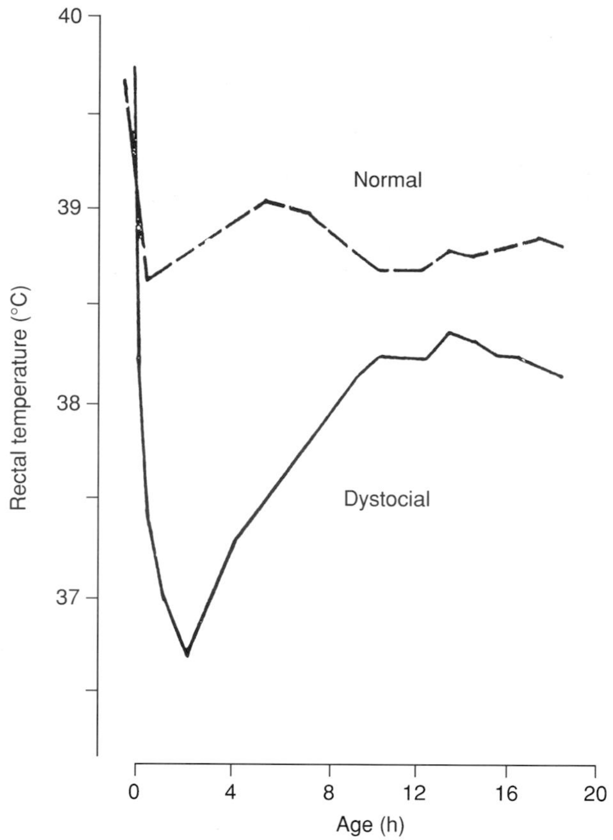

Fig. 7.13. Variation in rectal temperature of normal (eutocial) and dystocial calves in the first 20 h of life, when maintained at an ambient temperature of 10°C.

may not intuitively show these behaviours, especially after a difficult calving, and assistance may be necessary. In cold conditions, the calf may lack the strength to suckle, but fortunately, the time period over which immuno-globulins are absorbed from the gut is extended in such conditions.

Apart from air movement, other environmental factors will influence the susceptibility of cattle to temperature stress. On cloudless nights, outdoor cattle will lose most heat by radiation, and conversely on a sunny, cloudless day they gain a considerable amount of radiant heat. High humidity reduces the ability of cattle to lose heat by evaporative means, and any evaluation of the susceptibility of dairy cows to temperature stress should take account of this (Fig. 7.14).

The thermal properties of the floor are important, since this is the point of direct contact of the animal with the ground, to which cattle will lose heat by conduction. Bedding material in cubicles can act as an insulator as well as cushioning the impact of cows lying down and absorbing urine. In cold climates, such as in Canada, the concrete bases of cubicles may be laid over

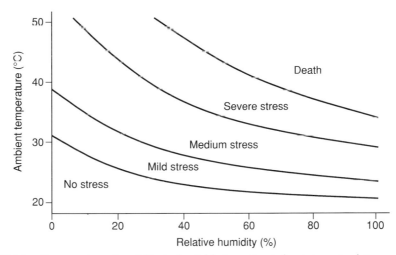

Fig. 7.14. Temperature–humidity index table to estimate heat stress in dairy cows. Relative humidity values are expressed as %.

an insulating layer and rubber matting used on top of the cubicle for extra insulation and comfort.

Water temperature is important in relation to the internal thermal environment of cattle. At high temperatures, careful positioning of supply pipes underground and insulating them above ground will provide cool drinking water for cattle. Running water pipes under the ceiling to the troughs will warm the water and exacerbate heat stress. In cold climates, pipes must be well insulated to ensure that the water supply does not freeze, which would cause a rapid reduction in milk production. Calves fed milk replacer at cold ambient temperatures have low intakes.

At high ambient temperatures, feed and water intake are adjusted to reduce heat stress. As temperature increases so does water intake, to replenish body water that is lost during sweating and by respiratory means. Feed intake decreases, particularly that of fibrous forages, often by about 10% to reduce the heat of fermentation in the rumen. It is preferable to feed a concentrated diet to cows at risk of heat stress and to include more sodium and potassium salts to replace those lost in sweating. In Israel, a diet of 70% concentrates and 30% roughages is fed to minimize the heat load to the animal. At the other extreme, at low ambient temperatures, failing to allow *ad libitum* feed intake can seriously reduce the ability of cattle to cope with the cold, as the cattle cannot increase their nutrient intake to cater for increased maintenance demands. Gut motility increases to allow for extra feed intake, which slightly reduces feed digestibility. Even in Britain, out-wintered cattle will eat more than those indoors to meet the extra demands for maintenance, but if adequate feed of reasonable quality is available performance will not be reduced.

Different cattle genotypes have different susceptibilities to temperature stress. In particular, the type of hair and its length will affect the insulatory value of body coverings. Usually, in hot climates, cattle develop short, shiny coats that reflect the sun's rays and transmit body heat effectively. The best coat colour is white for maximum reflection, but dark skin is valuable as the melanocytes protect against skin cancer. In cold climates, the coats of cattle become long and dull-looking, reducing heat reflection and maximizing the insulatory value and protection from the rain.

Other morphological features have developed over generations to combat temperature stress. In hot climates, cattle have developed long legs and loose skin, which increases the surface area from which heat can be lost. *Bos indicus* cattle are the most extreme. They have pronounced dewlaps and preputial sheaths, large ears and long legs. Fat stores are concentrated into a hump to avoid having a thick layer of subcutaneous fat that would restrict heat losses. *Bos indicus* cattle are predominantly white, but the variety of colorations of cattle in hot climates suggests that this is not too important. Apart from affecting heat loss, coat colour may influence the ability of insect parasites to locate a new host. In males, the scrotum is much extended to ensure that the testes are held as far away from the body heat as possible, while still being shaded from direct heat by the animal's torso. Spermatogenesis is impaired if conducted at core body temperature. In the female, oestrus is short and usually restricted to cooler parts of the day. As a result, it is important to look for cows in oestrus at these times, which is best achieved by overnight observation if labour is available. Feeding may also be concentrated into the cooler parts of the day, particularly that of grazing cows that are exposed to extreme heat without shade. The UCT of dairy cows is approximately 26°C, and less for high-yielding cows. It is therefore not surprising that even in 'cool' temperate countries such as Britain, cows can be seen seeking shade in summer months. The circadian variation in temperature is important, as in arid areas the hot days are often counterbalanced by cold nights, when the cattle lose the heat gained during the day. Heat stress also can cause embryo loss and, combined with the difficulties in observing oestrus, result in extended calving intervals, often well over 400 days. Gestation length is reduced by hot conditions, resulting in small calves with few brown adipose tissue reserves for the neonatal period.

If shade is provided, the roof is normally orientated east–west so that more ground is permanently shaded and therefore cool for the cattle to lie on. If the roof is orientated east–west, there is a greater movement of the shade provided because of the sun's movement, but the only benefit of this is to keep the lying area clean and dry. Regular movement of the shade may be possible in some small-scale installations to help to avoid any badly soiled areas. This is especially important for dairy cows, which are at risk of environmental mastitis if they lie in dirty conditions. Stocking densities in heat-stressed conditions should be low to maximize the air turbulence around each animal; 12–15 m² per animal is best for lactating cows. Free lying areas, rather than cubicles, are used so that each cow can have plenty of free air space around it.

A small area of hard-standing, say 3 m² per cow, is usually provided to allow cows in oestrus a good footing to mount each other.

Nowadays, the material to provide the shade is often synthetic, such as aluminium or asbestos-type boards. However, the latter are good conductors of heat, especially if the roof is low, and can even create a hotter temperature inside than the ambient temperature, albeit with no radiant heat load from the sun. Natural materials, such as straw and leaves, are better as they tend to be poor transmitters, but they provide a suitable environment for vermin to live in. If artificial materials are used, they must be painted white to reflect the heat. If the roof is low a double skin can provide a better heat protection than a single layer, but light, tall structures with aluminium profile sheets and no central supports are best. If the roof is tall, the airflow underneath it is unimpeded. Offsetting the two spans of the roof so that one is higher than the other at a central ridge will encourage airflow by the Venturi effect.

In extreme conditions, shade may be combined with sprinklers, in the collecting yard of lactating cows for example, or in the feeding passage. The improved comfort of the cows if they are sprayed before milking will help to encourage effective milk let-down. Cows can also be brought into the collecting yard on one or two other occasions per day to be cooled. Sprinklers can be sited both over the cows' heads and at floor level, the latter also serving to wash the cows' udders before milking. Droplet size is important. If it is too small, it will evaporate from the surface of the hair rather than penetrating through to the cow's skin, but if it is too large, it does not cover a big enough area of the animal's body and may run off onto the floor. Alternatively, misters may be used in the housed area for the dairy cows, but adequate drainage must be provided, and fans may be installed to reduce the heat load further by increasing the heat of evaporation from the animals. The misters are usually installed over the feeding passage, and cows will then be encouraged to feed when the misters are working. Often this would be after milking to stop the cows lying down and possibly contracting mastitis.

A particular problem in hot conditions with dairy cows is reproductive failure. In Israel and the southern USA, conception rates decrease from the normal values of 50–60% in winter to 20–25% in summer if the cows are not adequately cooled. The reduction in conception rate is mainly as a result of hyperthermia in sensitive reproductive tissues, but also partly because of reduced feed intake and the consequent malnutrition, and the failure of cows in oestrus to exhibit the normal behavioural signs, at least during the daytime. Hyperthermia in the ovarian tissues reduces the supremacy of the first dominant follicle, with reduced oestradiol secretion. After ovulation, the luteal cells of the corpus luteum produce less progesterone. Low conception rates persist into autumn, after temperatures have declined, probably because preovulatory follicles in the ovary have been adversely affected during the summer. Several months after hot summer temperatures have subsided, the ability of the theca cells of the corpus luteum to produce oestradiol is still impaired, resulting in reduced steroidogenic capacity of the ovarian follicles.

To conclude, the ability of farmers to effectively manage dairy and beef cattle in some of the most extreme temperatures on the globe is testament both to man's ability to effectively modify the environment, and to the adaptability of cattle in their thermoregulation. Cattle are now profitably and successfully kept in the hottest parts of the world. In extreme cold regions, it is not because the cattle could not survive that limits their population, but the shortage of fodder, which often makes it uneconomic to keep them in large numbers.

Calf Housing

Neonatal calves are vulnerable to infectious diseases because of their poorly developed immune system, and later their transition from monogastric to ruminant digestion renders them susceptible to gut infections and nutritional disorders. The severity of both of these possible problems depends on the suitability of the housing system. Because of the risk of transmission of diseases, many calves are kept in individual pens until they are weaned off reconstituted milk at 5–7 weeks. A minimum pen size of 1.5 × 2 m should be provided up to 6 weeks of age, and each pen provided with two bucket holders (one for reconstituted milk powder and one for water) and a rack for a fibrous feed – straw or hay. The milk powder can also be fed from a high level bucket with a drinking nipple attached. Buckets should be washed daily and left upturned to dry: if they are not washed properly or dried, they may become a source of cross-contamination.

Adequate drainage of the bedded area is important to reduce the relative humidity of the air and disease transmission either directly from the bed to the calf or by aerosol infection. The surface of the bedding must be dry to prevent bacterial proliferation. Drainage is especially important for calves on a predominantly liquid diet, which produce large quantities of urine. A floor slope of 1 in 20 will allow adequate drainage, while still being comfortable for calves to lie on.

Often calf pens are placed outside in hutches, with a gap of about 2 m between calves, which minimizes pathogen transmission between calves. Hutches provide the calf with an indoor and outdoor area, with straw usually provided in both. In the EU, calves in individual pens must have direct visual and tactile contact with at least one other calf through the walls of their pen, to allow the calf the opportunity to socialize directly with other calves. They cannot be tethered except for a short period (1 h) in groups for feeding. Group housing is compulsory after 8 weeks of age, except where calves have to be removed for veterinary treatment. Minimum space allowances for group housed calves of different weight are legally prescribed:

<150 kg	1.5 m² per calf
150–199 kg	2 m² per calf
200 kg or more	3 m² per calf.

Adequate bedding must be provided – this is usually straw, but sand and wood shavings are also used.

After weaning, calves should not be housed with older stock that may transmit pathogens to them. Good ventilation and low relative humidity are both essential in maintaining a healthy environment. It is now recognized that these are more important for calf health than keeping the calf warm. A common problem is the enclosing of calves in a confined space in order to reduce heat loss, particularly during cold conditions. In fact, adequate ventilation is even more important at low temperatures because of the high humidity in the air. High humidity encourages pathogen transfer and causes condensation on the walls and ceiling of the building, which can make the bedding damp. Insulation of the ceiling or at least providing a double skin can be valuable in reducing condensation.

Stockpeople may dislike working in a cold environment, but for calves, the temperature within their building is usually more than their lower critical temperature, which ranges from 9°C at birth to 0°C at 4 weeks of age in buildings with minimal drafts. However, a balance must be struck between good airflow through a building and draughty conditions at the level of the calves, which will severely weaken them.

Many calf houses rely on natural ventilation, providing the necessary six air changes per hour and 8–10 m³ of airspace per calf. This can be provided by a monopitch building or by building pens into the side of a double-span building (Fig. 7.15). In either case, care should be taken to avoid drafts at calf level by ensuring that there is a barrier to air that enters through the eaves cooling rapidly and falling onto the calves' backs. This can be a row of straw bales on top of the pens, or a solid board partition. Such a provision is most important for young calves and is particularly necessary in winter when the air cools more rapidly on entry to the building. Many calves are challenged by pneumonia when they are housed in poor housing with stale air and damp conditions. The dominant organisms are Para-influenza (PI) 3 and respiratory syncitial virus (RSV).

Housing Bulls

Dairy bulls are usually kept in an unnatural environment – solitary confinement. Cattle are social animals, and although bulls in semi-wild conditions have a greater inter-individual space than other classes of cattle, they still require companionship for good welfare. Isolation is part of the reason for the aggressive nature of many dairy bulls, but may be necessary for the protection of other cattle and humans on a dairy farm. The widespread use of artificial insemination is to be welcomed for this reason alone. If a bull has to be kept on a dairy farm, the housing must be secure and simple rules must be followed to ensure the safety of people working on the site. The animal must be free to roam within its accommodation box, which should be at least 3 m² with walls 1.5 m high. Some protection above this height is preferable to prevent

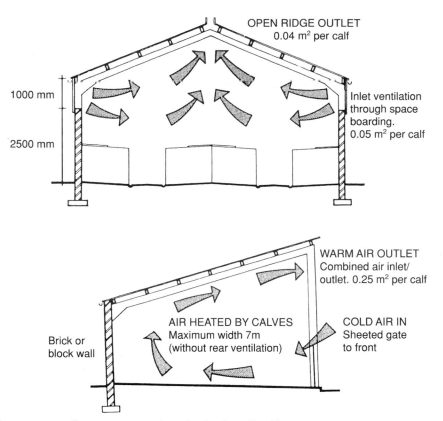

Fig. 7.15.   Airflow in monopitch and ridged roof buildings.

children or youths entering the pen. Warning signs should be posted on the outside walls or entry door. The bull should be fed and watered from outside, but regular, positive contact between bull and stockperson should be encouraged. The floor should be non-slip and care must be taken that the bulls feet do not get overgrown because of lack of exercise or too rich a diet.

The pen should be sited somewhere that the bull is likely to receive stimulation from passing cattle and/or humans. Positioning the pen adjacent to the collecting yard where cows wait to be milked is suitable as the bull will encourage cows to demonstrate oestrus in a place where the milker can record the cows easily. Adequate stimulation will avoid the bull masturbating to release tension. Regular handling will encourage the bull to relate well to people, but he quickly detects if the handler is fearful of him. It is better if the same person handles the bull, as a confident relationship should develop, but handlers should never tease bulls. Accidents are just as likely to happen during play as aggressive acts.

A service box should be next to the bull box so that the bull can be introduced easily without risk to the handler. Both the bull pen and service box should be equipped with escape gaps, 30 cm wide, to allow the handler rapid exit if necessary and at the corner of the service box there should be a pen where a cow can be introduced and led out in safety.

Cattle Marketing and Transport

Many cattle are still sold through live auction markets, although an increasing proportion go direct to an abattoir or are sold through an 'electronic auction', where the vendor and prospective purchasers do not meet, but communicate via the telephone or Internet. Some farmers are cooperating to send cattle in groups to abattoirs, thereby increasing their ability to control the price. The risk to a farmer of sending cattle direct to an abattoir is that if they achieve a low price, which may occur if they are given a low grade for the quality of the carcass, they cannot be returned to the farm for further fattening. Prices are set for the carcasses in abattoirs, compared with the live animal in auctions. If a producer sells to an abattoir, which then sells direct to a supermarket chain, the supply chain is shortened compared with selling through a market, and this assists in tracking the animal through the marketing chain. At auction markets, the vendors receive payment on the day of sale and the competitive bidding between purchasers may increase the price of the cattle. Purchasers buy the cattle per unit of weight and the bidders are given a guide to an animal's weight.

Electronic auctions prevent the stress to the cattle from being transported to and from live markets, and possible mistreatment in the market itself, but they do offer opportunities for purchasers to compete for the sale and for the buyer to withdraw the animal from sale if the price is inadequate. The difficulty is the grading of the animal, and some electronic auctions sell the animal subject to abattoir grading.

The welfare of cattle at markets may be compromised by two main factors. The first is the stress caused by the unnatural environment, the company of the unfamiliar animals and people and the disruption to their normal routine. Stress leads to the production of meat that is dark, firm and dry (DFD), particularly in bulls that are liable to fight when mixed. The second is the physical damage that may be caused by forcing cattle to move, particularly when this is done with the aid of electronic goads or sticks. When cattle slip or fall, bruising occurs that will reduce the value of the carcass.

New standards in markets are enabling the welfare of cattle to be improved. Cattle are less likely to be penned with strange animals and they are loaded and unloaded from lorries in an unhurried manner. Their physical requirements are catered for − non-slip floors, water provided in the pens, food and bedding provided if the cattle are accommodated overnight. In the UK, market personnel are instructed in how to move animals and sticks, etc.

are recommended for use only as extensions of the handler's arms, not to beat animals. Animals should not be moved by twisting their tail or, in the case of calves, lifting their back legs off the ground.

Moving cattle to and from markets or to an abattoir may result in a reduction in its value and poor welfare. Weight loss occurs during most journeys, which is mostly, but not all, fluids. During transport, cattle are subjected to noise, strange surroundings, odours and companions, over-crowding or sometimes isolation, hot or cold conditions, vibration and a lack of food and water. All contribute to stress and potential losses. However, this can be minimized by good practice. The vehicle should have good ventilation, with the exhaust fumes ducted well away from the animals. It should have good access for humans and an inspection light. The internal partitions should be adjustable and there should be no internal projections and adequate headroom.

When loading, the route to the lorry should be clear and preferably without shadows or pools of water. Cattle prefer to move from a dark area to light. Solid walls and curved raceways are best, but moveable gates are an acceptable alternative. The loading ramp should have side gates and a slope of less than 4 in 7 and cattle should not have to step up more than about 20 cm. The ramp should have battens for the cattle to get a grip and the floor of the ramp and the inside of the vehicle should be non-slip. Adequate staffing is essential for moving cattle and they should be moved slowly and preferably without the use of goads. Staff should be calm, confident and able to predict the animals' behaviour. Sticks can provide a useful means of extending the arm to accelerate movement, but should not be used to hit the animals with unnecessary force, using persuasion rather than aggression.

On the road, the vehicle should be driven with extra care, avoiding fast cornering and excessive braking. The animals should be checked on long journeys and in hot conditions a rest can be given by parking the vehicle in the shade. Once at the destination, it must be recognized that the cattle will be tired. A purpose-built bay should be used if possible to get them off, with the same characteristics as for loading. Providing access to food and water will help to reduce the stress of transport (Grandin, 2000).

Conclusions

Housed cattle are kept in an unnatural environment, but their welfare can be high if attention is given to meeting their needs. Different classes of cattle have different requirements for space, light etc., and it is imperative that stockpeople know the needs of animals within their care. Extremes of temperature must be allowed for and all animals within a group must be suitably accommodated, not just animals of average size. Particular care must be taken with isolated cattle, such as young calves or bulls, and when moving cattle.

References

Grandin, T. (ed.) (2000) *Livestock Handling and Transport*, 2nd edn. CAB International, Wallingford.

Heffner, H.E. and Heffner, R.S. (1993) Auditory perception. In: Phillips, C.J.C. and Piggins, D. (eds) *Farm Animals and the Environment.* CAB International, Wallingford, pp. 159–184.

Ingvartsen, K.L. and Andersen, H.R. (1993) Space allowance and type of housing for growing cattle – a review of performance and possible relation to neuroendocrine function. *Acta Agriculturae Scandinavica Section A – Animal Science* 43, 65–80.

Kuipers, A. and Rossing, W. (1996) Robotic milking of dairy cows. In: Phillips, C.J.C. (ed.) *Principles of Dairy Science.* CAB International, Wallingford, pp. 263–280.

Mein, G.A. (1992) Basic mechanics and testing of milking systems. In: Bramley, A.J., Dodd, F.H., Mein, G.A. and Bramley, J.A. (eds) *Machine Milking and Lactation.* Insight Books, Reading, pp. 235–284.

Metz, J.H.M. (1985) The reaction of cows to a short-term deprivation of lying. *Applied Animal Behavioural Science* 13, 301–307.

Phillips, C.J.C. (1990) Pedometric analysis of cattle locomotion. In: Murray, R.D. (ed.) *Update in Cattle Lameness, Proceedings of the VI International Symposium on Diseases of the Ruminant Digit.* British Cattle Veterinary Society, Liverpool, pp.163–176.

Phillips, C.J.C. (1993) *Cattle Behaviour.* Farming Press, Ipswich.

Phillips, C.J.C. and Schofield, S.A. (1990) The effect of environment and stage of the oestrus cycle on the behaviour of dairy cows. *Applied Animal Behaviour Science* 27, 21–31.

Ruckebusch, Y. (1972) The relevance of drowsiness in the circadian cycle of farm animals. *Animal Behaviour* 20, 637–643.

Stanford, C.F. (1991) *Health for the Farmer.* Farming Press, Ipswich.

Wathes, C.M. and Charles, D.R. (eds) (1994) *Livestock Housing.* CAB International, Wallingford.

Webster, A.J.F. (1984) *Calf Husbandry, Health and Welfare.* Granada Technical Books. London.

Whipp, J.I. (1992) Design and performance of milking parlours. In: Bramley, A.J., Dodd, F.H., Mein, G.A. and Bramley, J.A. (eds) *Machine Milking and Lactation.* Insight Books, Reading, pp. 273–310.

Further Reading

Bramley, A.J., Dodd, F., Mein, G. and Bramley, J. (eds) (1992) *Machine Milking and Lactation.* Insight Books, Reading.

Kelly, M. (1983) Good dairying housing design – a form of preventive housing? *Veterinary Record* 113, 582–586.

Phillips, C.J.C. and Piggins, D. (eds) (1992) *Farm Animals and the Environment.* CAB International, Wallingford.

University Federation for Animal Welfare (1994) *Management and Welfare of Farm Animals.* 3rd edn. UFAW, Wheathampstead.

Recommended Videos

Humane Slaughter Association (1989) *Livestock in Transit – Handle with Care.*
Humane Slaughter Association (1992) *To Market to Market.*

Cattle Growth and Rearing Systems

8

Introduction

Growth was once defined by the leading animal scientist Sir John Hammond as 'an increase in live weight until mature size'. Although this is a useful definition, it is not perfect as it could equally well apply to a cancerous tumour as to muscle growth. Scientists have defined it as 'cell enlargement and multiplication', but this is not easy or relevant for farmers to measure, or even more obscurely, 'an irreversible change over time in a measured dimension'. Cattle farmers are principally interested in 'an increase in saleable live weight until mature size'. The economist is primarily interested in the net value of the livestock and their potential to generate profits through sale of cattle products. From the point of view of producing a profit from rearing cattle for consumption, the critical statistic is the yield of lean meat, comprising carcass muscle and offal. Lean meat contains some 75% water, 18% protein, 3% non-protein nitrogen, 3% fat and 1% ash.

The conversion of food by cattle is inherently less efficient than that of monogastric animals because their digestive system utilizes a double digestion, an initial digestion by microbes followed by digestion of the microbial biomass and previously undigested feed by enzymes produced by the gastrointestinal tract of the cattle (Table 8.1). This complex system is necessary because of the low quality of the feed consumed by cattle, at least when they graze naturally. In addition, the energetic efficiency of suckled calves is low because of the need to maintain the mother as well as the offspring. However, the energy required post-slaughter for processing meat can be low, compared with many vegetable products. Foods from cereal grains and oilseeds can require considerable energy during processing and cooking, if for example, the grains are processed, the starch fermented into a loaf of bread, sliced and toasted before consumption. Any perceived inefficiency of land utilization for meat production must therefore consider the additional energy requirement for processing and cooking. If the energy requirement for production is evaluated relative to protein, rather than energy, the production of meat is more efficient than most

Table 8.1. The efficiency of feed energy utilization by meat producers (energy output as % of energy input).

Animal	Efficiency (%)
Pig	24
Chicken	14
Goat	8
Rabbit	8
Deer	8
Growing cattle	6
Suckler cattle	3
Suckled lamb	3

other feeds when processing costs are considered. Clearly, meat production cannot be dismissed as inefficient; it must be compared with other foods in terms of total resource use, and its requirements for different resources must be considered in the light of the scarcity of the resource in different regions.

An overriding principle of the growth of mammals is that their form is usually related to their function. Cattle were domesticated because of their suitable diet and reproduction, their ability to produce milk for human consumption and perhaps their temperament, but not necessarily their conformation. To enable them to digest coarse grasses, cattle have a large muscular abdomen containing the rumen. In addition, reflecting their polygynous breeding habits, the males have a large muscular neck and shoulders, to assist in competition for access to the females. They are not built for rapid movement and mountainous conditions, and hence they do not have well-developed limb and spinal muscles, unlike sheep and goats. Cattle therefore do not have the ideal muscle distribution for a meat producer that would favour large hind limbs, but they are well adapted to living off poor quality grasses. Recent breeding developments have gone some way to redress the balance, with double-muscled cattle having big, muscular hind limbs that are suitable for the efficient production of high-priced cuts of meat.

The growth of the body can be seen to undergo a series of waves of growth, with nervous tissue first, then bone, muscle and finally fat tissue. The initial wave of nervous tissue growth is essential to allow bodily functions to proceed, then bone is necessary to support muscle growth and finally fat tissue provides a store of excess energy intake that will be useful in periods of undernutrition. The waves of growth can be shortened by initially feeding cattle on a high plane of nutrition, accelerating their passage to the final wave of fat growth. Thus, animals on a high plane of nutrition end up with a high fat content at a given live weight. The ratio of bone:muscle is not affected by the plane of nutrition and is largely determined by the animal's physiological age. The final wave of growth, that of fat tissue accumulation, is particularly important for cattle that experience considerable variation in food quality between seasons. Fat acts, among other things, as a store of energy reserves that can be mobilized when little food is available.

Physically, these growth waves can be seen in the relative proportions of the different body parts. Calves have relatively large heads, because of the high content of nervous tissue. As the animal matures, the hindquarters become proportionately more significant, until finally the abdomen matures, providing a large rumen for microbial digestion of coarse grasses. The reduction in the proportion of the body as the head and skin can be seen as the dressing or killing-out percentage[1] increases as the animal grows (Table 8.2).

Because the growth slows down as the animal reaches mature weight, and more of the growth is fat tissue, which requires more energy intake than other tissues, the food conversion ratio[2] increases as cattle get older (Fig. 8.1). In fact, the food conversion ratio increases exponentially, making it important to slaughter at an early age to achieve an efficient use of food resources.

Early Growth

The growth of cattle, as of other animals, follows a sigmoidal or S-shaped curve. The initial constraint on growth rate is the development of the fetus, which must not grow so big that the mother has difficulty in giving birth. This can place a limitation on the size of sire that can be used to father a calf from a small cow. Ideally, the maintenance cost of the dam should be minimized by small size in relation to the sire, but if taken to extremes this can lead to an unacceptable level of calving difficulties.

The causes of calving difficulties are multifactorial, but genetic predisposition plays an important part. The proportion of Friesian heifers with calving difficulties increases from about 2% when the sire is from a small breed, such as the Aberdeen Angus or Hereford, to 8–10% when the sire is one of the large continental types, such as Limousin, Simmental or Charolais. With older cows there are fewer problems, and the size and breed of cow is important. Within breeds there is large variation between individual bulls, and this does not relate just to their body size. The shape and size of the head is particularly important, as this is often the most difficult part for the cow to expel through the birth canal.

Table 8.2. Changes in carcass weight and composition with increases in the live weight of steers.

Live weight (kg):	307	386	466	545
Slaughter weight (kg)	167	217	268	322
Dressing (%)	55	56	57	59
Bone (%)	18	16	15	14
Lean meat (%)	65	64	61	58
Fat (%)	14	18	24	29

1 Carcass weight (kg) as a proportion of live weight (kg).
2 Feed intake (kg) divided by live weight gain (kg).

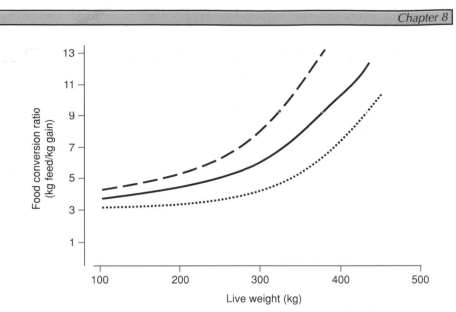

Fig. 8.1. The food conversion ratio of steers (——), heifers (– –), and bulls (⋯⋯) at different live weights.

Postnatal Growth

After birth, the calf enters an exponential phase of growth, providing it is fed high-quality nutrients. The main factors determining growth from this period to slaughter weight are:

- sex;
- hormones and hormone analogues;
- nutrition;
- genotype;
- climate;
- health.

We will consider here the effects of sex and hormonal moderation, the remaining factors having been considered in other chapters in this book.

Sex effects on growth and cattle management

Sex differences in cattle growth are mainly explained by differences in mature size. Three sex types are distinguishable – bulls, castrates (otherwise known as steers or bullocks) and heifers. Bulls have a low dressing percentage because of their large heads, as a result of which they have a greater proportion of fore-than hindquarter (Table 8.3). At any given slaughter weight they are a smaller proportion of their final weight than steers and even less than heifers, i.e. they are the least mature. As a result, they have the highest proportion of bone and muscle and lowest proportion of fat, compared with steers and heifers. A

Table 8.3. Carcass composition (%) of bulls, steers and heifers at 450-kg slaughter weight.

	Bulls	Steers	Heifers
Dressing	56	58	58
Hindquarter	47	50	50
Forequarter	53	50	50
Bone	16	15	13
Lean meat	68	58	54
Expensive muscle	53	54	56
Fat	13	25	31

slaughter weight of 450 kg would not normally be considered sufficient for bulls, but it would for heifers.

Cattle rearing methods are similar for the male and female animal, but the purpose may be quite different, because a much larger proportion of females must be retained for breeding compared with the males. Bulls are, therefore, almost all reared for meat, and may be castrated to improve their temperament or to reduce the time that they take to reach a certain level of fatness. Castration also prevents breeding, enabling only progeny-tested bulls to be used for breeding offspring with desirable characteristics. Castration alters meat quality, increasing fat content at a given weight and, because the stock are less likely to fight when stressed before slaughter, they are not prone to dark cutting. Dark cutting is caused by inadequate glycogen stores at slaughter, leading to a high ultimate pH of over 6 in meat. This leads to rapid deterioration, and although the meat is tender, the flavour is not pleasant. It is not confined to bulls, but is more common in these animals because of their excitability before slaughter, especially double-muscled bulls that are particularly stress susceptible. The frequency of dark cutting is about 3% for steers and heifers, 7% for cows and 12% for bulls, although it can be up to 20% in certain herds. The *M. longissimus dorsi* is the muscle most affected, but effects are observed throughout the carcass and the carcass value will be reduced by at least 25%.

The disadvantages of castrating bulls are, firstly, the decrease in food conversion efficiency, mainly because of a reduced growth rate, which increases the maintenance requirement as a proportion of total energy requirements; secondly, the increased fat content at a given carcass weight, and thirdly, the inability to grow cattle to a heavy weight to dilute the high cost of the calf in relation to its weight.

Castration is common in the UK, where cattle are usually grazed during summer because of the favourable grass growing condition. As many of the fields contain footpaths that are public rights of way, castration is often performed purely for safety reasons, usually at approximately 3 months of age. It can be painful to the calf and adequate anaesthesia and analgesia should be used.

The increasing requirements of the public for lean meat have encouraged farmers to consider leaving their male cattle entire. However, farmers should consider whether a field is likely to be visited by groups of walkers with dogs, children on day-trips or lone walkers, all of which might attract the attention of bulls, or even charging by a group of young heifers. If there are young calves with their mothers, this will encourage defensive action by the cows. Farmers with a right of way through their fields may wish to fence it off to protect the public from possible attack by the cattle. In the UK, bulls of recognized dairy breeds (Ayrshire, Friesian, Holstein, Dairy Shorthorn, Guernsey, Jersey and Kerry) cannot legally be grazed in fields with a public right of way, and bulls from a beef breed can only be grazed if cows or heifers are present. The 'fields' specified in UK law do not include unfenced grazing, such as moorland or open fell. If bulls are kept in fields, farmers should ensure that the fences, gates, etc., are in good condition and that paths are clearly marked and signs erected warning members of the public that there is a bull in the field.

When bulls are handled by farm workers, there is also a risk of injury, against which precautions should be taken. A 5-cm diameter ring should be inserted in the nose when a bull reaches 10 months of age to enable him to be safely led. The bull should learn to associate human contact with pleasurable events, such as feeding. At least two people should be present whenever a bull is handled and they should work from a mobile sanctuary, such as a farm vehicle, whenever a bull is handled in a field.

Female cattle may be reared for replacement dairy cows, replacement suckler cows or for meat. They start to put on fat tissue before castrates or bulls. Heifers grow to a smaller mature size than steers, which are in turn smaller than bulls. They also mature earlier, which means that they must be slaughtered for meat at a younger and lighter age than steers, which must also be younger and lighter than bulls, if a similar level of fatness is required in the carcass. Because the heifers lay down tissue with a greater fat content than steers or bulls at a given live weight, their food conversion efficiency is less (Fig. 8.1).

Hormones and hormone analogues to modify growth and lactation

The sex effects on growth are driven by the hormonal complement of the sexes. Most potent are the androgens, principally testosterone, produced predominantly in the testes, and important in increasing the efficiency of growth by increasing the nitrogen incorporation into muscles. This is probably regulated by growth hormone. Androgens also cause epiphyseal plate fusion in bones and exogenously administered androgens can reduce skeletal size. Exogenous androgens such as trenbalone acetate, if permitted, have their greatest effect in heifers or cull cows, because of the low level of natural male steroids in the female. In the EU, both synthetic and naturally occurring growth promoting hormones were banned in 1986. Oestrogens are also potent growth stimulators in young steers. They increase growth hormone, leading to increased muscle production, decreased fat production and reduced losses of urinary nitrogen. In the older animal, oestrogens cause epiphyseal plate fusion in bones in the same way as androgens. Both synthetic oestrogen-mimicking agents, such as diethylstilboestrol and zeranol, and naturally occurring female steroids, principally oestradiol, are most effective on steers, although combined action trenbalone acetate and oestradiol implants are effective in stimulating growth in bulls, steers and calves. The efficacy of synthetic steroid use is greatest in cattle, intermediate in sheep and of limited value in pigs. Because of the stimulation of muscle growth by both oestrogenic and androgenic hormones or hormone mimicking agents, it is often necessary to supply extra rumen-undegradable protein to implanted animals.

Other hormone mediators of growth include the β-agonists, which are synthetic analogues of adrenalin and noradrenalin, such as clenbuterol and cimaterol. These reduce intramuscular fat considerably, by up to 30%, with a corresponding increase in protein deposition of 10–15%. As a result, the food conversion efficiency is often increased by a similar proportion, 10–15%. The effects on weight gain and feed intake are variable, depending on the relative impact on fat and protein deposition. There is evidence that cattle treated with β-agonists are more susceptible to dark cutting, and the low level of muscle glycogen and carcass fat can give rise to cold shortening (cross-bonding between actin and myosin fibres) if the carcass is rapidly chilled *post mortem* to 10–15°C. The increase in carcass yield may be accompanied by smaller non-carcass components. The action of β-agonists is not sex specific but all animals are susceptible to tachycardia (elevated heart rate) and increased basal metabolism rate, which may be perceived as reducing their welfare. The risk of residues is low as the β-agonists are rapidly metabolized, and after withdrawal of the substance from the feed the animal's nitrogen metabolism rapidly reverts to normal.

The growth hormone complex itself can be directly moderated to influence both milk production and body growth. Bovine somatotrophin, or bST, has been well documented to increase the milk production in cattle by up to 25%. This effect was first reported for growth hormone in the early part of the 20th century, but the technology remained dormant until the growth hormone analogue bST came to be mass produced by recombinant DNA techniques in the second half of the century. The galactopoietic effect, and in particular the increase in milk fat and lactose production, is caused by the lipolytic effect of bST. However, despite the obvious increases in milk production efficiency in cows injected with bST, there remains public and some scientific concern that its use is unjustified. This derives from possible adverse effects on the welfare of the cows and effects on the health of humans consuming milk from treated cows. The welfare impact of bST is in almost all respects undeniably negative, but then so are the effects of breeding cows for increased milk yields. There is particular concern that bST increases the prevalence of those diseases associated with high milk yields in the cow, principally mastitis and reproductive failure. Cows are undoubtedly more likely to get mastitis when bST is administered, but the greater risk is only as much as would be predicted by the increase in milk yield, and perhaps less if the increase in insulin-like growth factor 1 (IGF-1) that accompanies bST administration has immunological benefit. Some of the adverse effects of high yields after bST administration are offset by the eventual increase in voluntary feed intake. The cow's homeostatic mechanisms still function, so that if a cow is injected with bST and no additional food is available, the increase in milk yield is negligible or very small, depending on the level of body reserves that she has. For this reason, bST has little role in most situations in developing countries because of the low quality and sometimes quantity of food available for feeding cows. Certainly, the effective use of bST is only possible where the farmer's management is good. Some potentially negative effects can be avoided by good management, for example, bST administration can be delayed after calving until a cow is pregnant, avoiding the difficulty in achieving conception in a cow in a negative energy balance.

Another potentially serious problem with bST is that the fortnightly injections into the rump can cause serious localized reactions, which are caused by the injectate itself not the injection. However, the abscesses quickly regress and subcutaneous injection rather than intramuscular is now recommended to minimize the local impact of the chemical. Another potentially adverse effect of bST is disproportionate effects on certain organs. In just the same way that certain growth promoters do not affect all parts of the body equally, so bST administration has been shown to increase glomerular hypertrophy in the kidney. This could have implications for the risk of kidney failure, although sufficient evidence is yet to be accumulated.

In relation to the possible effects on human consumers, the increase in IGF-1 content is of concern, as this moderates a range of processes in man (and the cow), and has been shown to cause intestinal cell proliferation in rodents at least. However, it is unclear whether the oral consumption of milk with high concentrations of IGF-1 will result in increased circulating IGF-1 in humans.

Surprisingly the administration of growth hormone to growing cattle does not result in large increases in muscle growth, perhaps because of the lack of additional receptors. However, immunization against the agonist of bST, somatostatin, can increase growth, but it also tends to increase carcass fatness. Somatostatin also inhibits other hormones, such as insulin and the thyroid hormones, which may explain its action.

Other growth promoters

Antimicrobial compounds are routinely used in some cattle production systems to modify the gut microflora. The most commonly used is monensin sodium, which was originally developed as a coccidiostat for poultry. In the rumen of cattle it is active in reducing the population of acetate and hydrogen producing bacteria, such as *Ruminococcus* species and *Butyrivibrio fibrisolvens*, allowing propionate producers such as *Selenemonas ruminatum* to flourish. This increases the efficiency of growth by about 5%, partly because acetate production is accompanied by methane loss via eructation. As a result of its mode of action, there are no effects of such growth promoters on carcass composition. The widespread use of monensin sodium was not possible until it could be incorporated into feed blocks that could be offered to the cattle when they were out at pasture. If cattle are offered feeds with added monensin sodium indoors and then turned out to pasture with no supplement, there is a considerable check to growth as the rumen microflora adapts.

There is increasing concern over the routine use of antimicrobial compounds in cattle production systems, principally because of the risk of transfer of resistant bacteria from animals to humans via the food chain and the possible transfer of resistant genes from animal bacteria to human pathogens. Currently within the EU, animal feed additives are only allowed if there is no known adverse effect on human or animal health, or the environment. Although there were originally ten licensed antimicrobial growth promoters, four of these (bacitracin zinc, spiramycin, tylosin phosphate and virginiamycin) were withdrawn in 1999 because of fears that human health would ultimately be compromised by their use. A human antibiotic similar to virginiamycin is currently being developed, which it is suspected could be rendered ineffective with continued use of virginiamycin in animal food. A further two antibiotics (olaquindox and carbadox) have been banned because of possible risks to human health during the manufacturing process, leaving only four that can legally be used (monensin sodium, salinomycin sodium, avilamycin and flavophospholipol).

Probiotics are an alternative to antibiotics when they are used therapeutically, but are not an alternative growth promoter. They promote colonization of the gut by benign bacteria, such as *Lactobacilli*, thereby excluding pathogenic bacteria by reducing nutrient availability or, in the case of *Lactobacilli*, acidifying the gut contents with lactic acid. Their use in cattle is restricted to the pre-ruminant calf, where they may prevent *E. coli* from colonizing the gut and causing scours.

Photoperiodic manipulation of cattle growth can achieve desirable changes in composition, although it is doubtful whether weight gain is increased. In autumn, cattle in natural photoperiod naturally begin to divert nutrients from muscle to fat deposition, to give them a store of nutrients that can sustain them through the winter. This would have been of particular benefit to wild cattle, although nowadays adequate conserved food is usually made available to prevent cattle losing weight in winter. Many wild herbivores naturally lose weight in winter; for example, the bison, a close relative of cattle, catabolizes considerable amounts of fat tissue through the winter on the American plains. In intensive rearing of cattle, food is available in similar quantity and quality throughout the year, but cattle still use the cue of declining photoperiod to start diverting more nutrients to fat deposition in autumn. By extending the photoperiod in autumn to 16 h of light daily, cattle metabolism can be altered to deposit lean tissue as if they were still in summer. This could be useful if they are to be slaughtered mid-winter, as they will put on more muscle and less fat tissue. If, however, they are being kept until the spring, photoperiodic manipulation will have no benefit, as cattle in natural photoperiod start to divert nutrients away from fat deposition to muscle growth in spring.

Compensatory Growth

An animal whose growth has been retarded on a low plane of nutrition will exhibit, when returned to a high plane of nutrition, a faster growth than would normally be expected. This phenomenon of compensatory growth is well developed in cattle, because they evolved in situations where there would be considerable fluctuation in their food supply between seasons. It enables farmers to use food resources optimally, taking into account fluctuations in availability and the price of different food resources.

Cattle who undergo a period of growth retardation, followed by compensatory growth, may experience some delay in reaching maturity (Fig. 8.2), particularly if the restriction is imposed for a long time and is severe. If the restriction is imposed at a very early age, full compensation may never occur. The evidence for this is stronger in pigs, where the existence of permanently stunted animals (runts) is quite common. In cattle, the greater the restriction, the faster they will compensate. Periods of restriction are characterized by a reduction in maintenance requirements, partly because of reduced size and activity of some vital organs, principally the liver and gastrointestinal tract, and perhaps partly because of reduced activity (although food searching

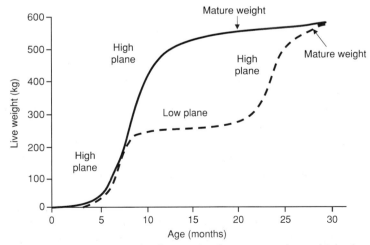

Fig. 8.2. The comparative growth of two animals, one reared on a high plane of nutrition throughout (———), the other transferred from a high plane to a low plane and then returned to a high plane (– –), demonstrating differences in the time at which maturity is reached.

will increase in some conditions). The compensation period is characterized by increased voluntary feed intake, leading to a rapid increase in weight gain, although some of the initial weight gain may be caused by increased gut fill and restoration of vital organ weight. Adequate food availability is essential for cattle to compensate. If they are deprived of sufficient nutrients in winter, as beef cattle may be, they can only catch up at pasture if there is sufficient herbage available for high intakes, i.e. a herbage height of 8–10 cm or more. The food that is provided during realimentation should be of high quality, so low-quality straw will not allow cattle to catch up if they have been restricted.

During the restriction period, the growth of the tissue that is most actively maturing is most reduced. Less fat deposition around the gastrointestinal tract is commonly observed, allowing feed intake to increase rapidly when adequate high-quality food becomes available. The reduction in lipogenesis that occurs during the restriction may prevent the food conversion efficiency being reduced much, because fewer nutrients are required to deposit muscle tissue than fat. If young animals are restricted, they may delay bone maturation to allow this to occur when food is more plentiful. During realimentation, the high growth rates, relative to body size, allow the food conversion ratio to be reduced, compared with animals of the same chronological age. This can be used to good effect by farmers buying store cattle in spring, who are well aware that they will grow rapidly and efficiently when turned out to pasture. For this reason, the price of store cattle usually increases quite substantially just before the time when cattle can be turned out to pasture in spring. Over their entire lifetime, however, cattle that have been restricted and then realimentated will have a reduced food conversion efficiency compared with cattle that have been on a high plane of nutrition throughout.

Measuring Growth and Body Condition

An efficient cattle-rearing industry cannot exist without accurate measures of growth, as the animals should be slaughtered at the optimum time in relation to their weight and composition and with due regard to feed availability and cost. Regular monitoring of growth throughout the rearing period will enable feeding schedules to be adjusted so that growth targets can be met. The methods used by farmers to determine when to alter their feeding regime and when to send the animals to slaughter are currently quite simple, although more sophisticated methods are being developed.

An important pre-requisite for monitoring growth and body condition is to be able to identify all animals in a herd accurately. This can be achieved by ear tags, although these are prone to being torn out, ankle straps, which get dirty and difficult to read, neck collars and subcutaneous electronic implants. Ankle straps and neck collars may contain electronic chips to relay information automatically to a receiver, which can send the information to a central computer for processing. In the aftermath of the BSE crisis in the UK, a cattle passport system was established in the EU to allow traceability of all cattle. Passports have to be applied for by 15 days after birth and will accompany the animal whenever it is sold or goes to slaughter. Movement labels with barcodes allow animals to be tracked quickly, and eventually electronic tagging will allow animals to be tracked from birth to their final destination, i.e. the retailer. Further details of the British cattle passport system are given in Chapter 6.

Measuring growth by weighing the animal will give limited information, because of fluctuations in gut fill, which comprises up to 25% of total weight in the adult animal. The rapid passage of food through the gastrointestinal tract is accompanied by a reduction in the animal's 'live weight', sometimes by up to 40 kg when a mature animal is turned out to pasture in spring. Weighing scales have improved in accuracy in recent years and most are now electronic, which take mean readings of repeated measurements from load cells while the animal is standing in a crate. Scales based on a lever connected to a spring balance are prone to error because of animal movement in the crate. Some dairy parlours are fitted with electronic scales that record each cow's weight as she leaves the parlour, and can alert the herdsperson if there is a major change in the herd's mean weight or the weight of an individual animal.

Visual assessment is used to estimate body muscle and fat cover in both the dairy and beef industries. The measurement of the condition of cattle was first described in 1917, where it was used to estimate the ratio of fat to non-fat composition of cattle, but it was not widely used until Lowman *et al.* (1976) proposed a five-point scoring method based on palpation of the spinous processes and the tailhead that had already been successfully developed for sheep. At this time, intensification of the dairy industry necessitated better assessment and regulation of body fat reserves. Nowadays the herdsperson should regularly check the 'body condition' of the cows by either a visual assessment from behind each cow, together with palpation of the spinal chord and the tuber coxae/tuber ischii region if possible (Fig. 8.3). Assigning scores of 1–5, depending on fat, and to a lesser extent muscle cover, it will enable the herdsperson to put the herd's body condition scores in the context of optimum scores at certain stages of the lactation cycle that have determined by research. In Australia, an eight-point system was also developed in the 1970s (Earle, 1976) and in New Zealand a ten-point system is used (Scott *et al.*, 1980). Optimum scores are considered in Chapter 2. It is a subjective measurement and differences in an animal's frame size and shape can lead to different body condition scores being attributed to an animal despite there being similar levels of fat/muscle cover. Nevertheless, it is possible to attain a reasonable degree of uniformity between people scoring the cattle, which enables the system to be used for advisory and research purposes.

Beef cattle are also subject to visual appraisal, but mainly *post mortem*. In Europe, cattle are graded for conformation, which is a visual assessment of the shape of a carcass, in particular whether it is endomorphic or ectomorphic.

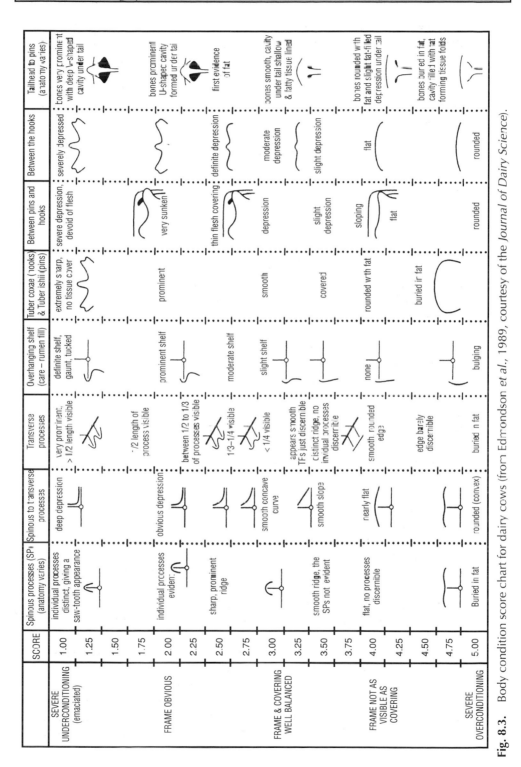

Fig. 8.3. Body condition score chart for dairy cows (from Edmondson *et al.*, 1989, courtesy of the *Journal of Dairy Science*).

This mainly reflects the muscle and fat cover. Carcasses are also specifically graded on a visual appraisal of fat cover.

The most important indicators of the saleable beef content of a carcass that are recorded in an abattoir are the weight, fat cover, conformation and breed, in that order. The killing-out percentage is also important and is influenced by the animal's age, sex and breed. The grades awarded in an abattoir influence the price paid for the carcass, and it enables retailers to specify the type of carcass that they wish to purchase, according to the market that they supply. Although the grading is determined visually, ultrasonic probes are now available for a more objective appraisal. In the live animal, the use of ultrasonic scanning to determine the optimum time of slaughter is made difficult by the hair covering, which prevents a good contact being made with the skin.

Cattle Rearing Systems

Suckled calf production

Producing suckled calves for beef is inherently inefficient because the calf is the main product and the maintenance cost of keeping a cow has to be included in any consideration of efficiency. In contrast, calves from a dairy herd are a by-product of the milk production industry, and the maintenance of the dairy cow is covered by the milk output of the cow. Hence in most parts of the world suckler cows are confined to land that can be used for few other agricultural operations. The fact that suckler cow systems survive is as a result of their ability to use low-grade land and the high quality of the product, which meets a need for best beef. In areas where the land used is of low fertility, or the climate is hostile, as in central Australia, or the terrain is too hilly for cultivation on a large scale, as in the uplands of Great Britain, suckler cows can be kept in a low-input system to produce store calves or young cattle that can be transported to more favourable regions for rearing to finishing weight on better quality diets. This can be achieved on lowland pastures or, in the Americas, in feedlots that contain up to 75,000 cattle each. Mountainous regions often have too little forage production potential for efficient suckled calf production, and the high nutritive requirements of a cow are much greater than a sheep or goat. Hence, in the UK there is about one cow for every five sheep in the uplands, but only one cow for every 20 sheep in the hill areas. A small number of cattle are beneficial on most sheep farms as the unselective grazing habit of cattle will keep any rough grass in check, whereas sheep will select only young grass tillers and may allow coarse, unproductive grasses to grow unhindered in summer. A harsh climate does not in itself prevent suckler cows being kept and cattle are agile enough to cope with foraging on steep slopes, but they must have adequate forage to provide for their maintenance requirements. In dryland farming systems, where the constraint of low forage availability equally applies, the introduction of fencing for rotational grazing systems has enabled pasture to be more effectively rationed, preventing overgrazing.

As there is no seasonal anoestrus in cattle, farmers can keep a suckler herd that calves at any time of year. The most favoured calving period for herds kept in harsh hill conditions is in spring, as peak nutrient requirements for cow and calf coincide with good availability of grass in summer. In upland regions cows are most likely to calve in the autumn. Since most calves are sold in autumn sales, an autumn-born calf aged 10–12 months will be larger and attract a higher price in the sales than a spring-born calf, sold at 4–6 months of age. However, the winter-feed requirements are much greater for an autumn-calving cow. Any underfed cows will not produce sufficient milk to sustain rapid growth in their calves, and they will be difficult to get back in calf. The cow usually replenishes weight lost during winter when she is at pasture in the summer, but calf growth will suffer if inadequate feed is available in winter. Typically, a medium-sized suckler cow will have an energy requirement of about 100 MJ ME day^{-1} during winter. This translates into about 8 t of silage for an autumn-calving cow, whereas the requirements of a spring-calving cow are likely to only be about two-thirds of this amount. This means that more land must be reserved for forage conservation for autumn-calvers, perhaps 60% of the grassland area for two cuts, compared with perhaps 40% for a spring-calving herd. In many hill farms, setting such a high proportion of land aside for conservation, when the grass growing season is short anyway, is not possible because of constraints of the terrain and the need for grazing. The introduction of big-bale silage making and handling machines has assisted many farms in moving from haymaking, with all its difficulties in wet areas, to conserving their fodder as silage.

A severe constraint to efficient suckler cow management is the difficulty in getting cows in calf. The non-cycling period is often 50–60 days, compared with only about 25 days in dairy cattle. Much of this is down to lactational anoestrus, which arises from a psychological inhibition of luteinizing hormone secretion when the calf is near the cow. Restriction of the suckling period to once or, at most, twice daily allows ovulation to proceed naturally, but is impractical for many farms in developed countries. It is widely practised in developing countries, especially where *B. indicus* cows are used, as they require the presence of the calf to let down their milk for extraction by machine. Calves may be allowed to suckle their mothers for a period of about 20 min at each milking and the residual milk, which has a high fat content, is then collected for human consumption. Alternatively, with *B. taurus* cows, the calf can be allowed to suckle after milk for human consumption has been taken. This has the advantage that the calf is able to extract milk by bunting the udder that could not easily be extracted by machine. It is, however, rather too rich in fat content for the calf and may cause scouring. Also, the herdsperson may be encouraged to take too much milk for sale, not allowing enough for the calves. The increase in milk yields by this technique increases the nutrient demands of the cow, which if not met will make it unlikely that she can be rebred, even if lactational anoestrus is prevented by restricting suckling to once or twice daily.

Another cause of lactational anoestrus in spring-calving cows is the increasing photoperiod at the time at which they would normally be bred. Cows, although not strongly seasonal in their reproductive cycle, do show longer anoestrus in early summer, because if they conceived then the calf would be borne in mid-winter when there was little food available.

Conception rate in suckler cows is typically 50–60%. This would be greater if cows were in better condition at service, particularly those that calve in autumn. Because of the difficulties in providing adequate forage for suckler cows in winter, target body condition scores are usually less in late winter/ early spring for autumn-calvers than for spring calvers in autumn which are at the same stage in the pregnancy cycle (Table 8.4). An autumn-calver is expected to calve in better condition to allow for this. Most spring-calvers meet their targets but many autumn-calvers are in inadequate condition at mating and mid-pregnancy. Cows that are served when their body condition is more than 2.5 are likely to experience a high incidence of calving difficulty. This may occur with summer calvers that have plenty of grass before and after calving. However, summer calving in productive upland or lowland farms fits in well with sheep production, as the cows have their greatest nutrient demand when the lambs are being weaned in mid–late summer.

The objective when rearing suckled calves is usually to produce a single weaned calf per year. In some more intensive operations, particularly on lowland farms, additional calves may be purchased and given to the cow to suckle. Recent emphasis on ensuring that suckler cows are produced from high-yielding breeds, such as Friesian crosses, allows the cow to suckle more than one calf, but increases the cow's nutrient requirements.

The breeds used for suckled calf production are many, and breed diversity has been preserved better than in the dairy industry. The classic beef breeds, such as the Hereford, Aberdeen Angus and British Shorthorn, still have their followers, often selling their product as superior quality to beef from the newer breeds. These classic breeds have characteristic small endomorphic carcasses that were largely developed for intensive beef production in the last century. Recently, there have been efforts to increase the size to meet modern demands, particularly in the Hereford. In the hills a range of breeds was developed that are particularly hardy, notably the Highland, the Galloway and

Table 8.4. Target condition scores for autumn- and spring-calving cows (courtesy of the British Society of Animal Science).

	Serving	Mid-pregnancy	Calving
Autumn-calver	2.5	2	3
Spring-calver	2.5	3	2.5

Belted Galloway, the Welsh Black (a former dual-purpose breed) and the Luing. The latter was produced as Beef Shorthorn × Highland cattle on the Western Isles of Scotland. Modern practice is often to cross a Continental breed, particularly the Charolais, or a classic beef breed, such as the Hereford or the Aberdeen Angus with the Friesian, which ensures a high milk yield for rearing good calves and allows the overheads of one-half of the cross at least to be covered by the dairy herd, and the benefits of hybrid vigour arising from breed complementarity to be realized. In the case of the Charolais-cross, the maintenance requirement may well be increased by the larger size of the cow compared with the traditional breeds. This is not compensated fully by increased calf weight and growth potential, leading to a reduction in the efficiency of production.

Some improvement in efficiency can be achieved if high-yielding suckler cows are allowed to suckle two or more calves. This may be controlled by the stockperson, so that the calves are brought to the cows and removed after suckling. With an average herd size in the UK of just 18 cows, compared with about 70 for dairy herds, this may be feasible using existing farm labour on beef farms. Each calf will require about 5 l of milk daily in two feeds per day, if performance is not to suffer. Alternatively, the cows and calves can be grazed together, where the calves take their chances with whatever cow will allow them to suckle. In this system, some cows may reject calves and some calves will be reluctant to suckle, leading to a less predictable but lower input system.

Instead of giving cows surplus calves, usually from the dairy industry, cows can be hormonally induced to have twins. There is a natural incidence of twinning of about 2%, but this is not sufficient for selecting a herd of twin-bearing cows. The disadvantages of twins are the small size of the calves (about 75% of a single) with the possibility of permanently stunting the calves' growth, the increased calf mortality (up to 10%), the risk of freemartins and the increased nutrient requirements of the cow. This long list explains why twinning is not welcomed by most suckler cow producers, who aim to produce one calf from each cow. A better alternative for the farm to increase the profit from its operations is to rear the calves on to slaughter weight. This entails having several groups of cattle on the farm, but the producer can then reap the reward of producing high-quality cattle for the premium market. He may also be able to capitalize on the several quality assurance schemes that are now popular (see Chapter 6). Some premium has always been available to producers of suckled calves, as buyers will pay appreciably more for calves that have been reared on their mother's milk. Such calves are recognized as having better health and having grown faster on the milk and grass combination that is in many ways nutritionally ideal. This rearing method is identifiable to buyers in the form of the 'suckler bloom', a shiny coat and bright eyes, indicating that the nutrition and health of the calves has been of a high standard.

One alternative to suckling cows that has often been promoted on the grounds of efficiency, but is not popular with the farmers, is the once-bred heifer. In this system, heifers are reared and produce one calf, after which they are slaughtered for beef, the calf then entering the system. In theory, this is a good way to increase the production of beef and not dairy products, when the latter are in surplus. However, it is largely unsuccessful because farmers are reluctant to slaughter heifers that they have put much effort into rearing and the meat is not of the quality demanded by the industry.

Finishing store cattle

Store cattle are usually finished on lowland farms in the UK, and there is a large range of cattle of different ages and weights available, so the farmers have to be able to adjust their systems to fit the type of cattle that they have. The change of farm and the transport stress the animals and enzootic pneumonia or 'transit fever' is common. Affected cattle must be rapidly treated with a broad-spectrum antibiotic. Store cattle can be fed a variety of diets, but any change in diet should be introduced gradually. High-quality forages, especially maize silage, root crops and cereals are most likely to be included in the ration, but waste products from the vegetable industry, such as stock feed potatoes, can be included and reduce the cost of the ration. The skill of the farmer in buying low-cost feeds, and cattle, undoubtedly plays a part in the profitability of the store finishing enterprise.

The cattle can be finished indoors in winter, in which case they are usually fed good-quality forage and a limited amount of concentrates, perhaps 2–3 kg day^{-1} per head. Alternatively, they can be finished at pasture. If the cattle are purchased early or mid-winter this will only be applicable to animals of late maturing breeds. In this case, they should not be fed a high-quality ration in the winter or the cost of finishing will be too great: silage alone or clean straw and a small amount of concentrates ($c.$ 1.5 kg day^{-1} per head) would be appropriate. They may only grow at about 0.5 kg day^{-1} during winter but they will compensate for this when they are at pasture. The grazing cattle can be sold when they are finished or, in an emergency when the grass availability is very low. However, grass supply will influence the price of cattle and it is generally best to follow guidelines for good management of grazing (see Chapter 4) and finish the cattle at the target weight, rather than selling them early.

In America, the systems of finishing store cattle are less variable, mainly because there is a large supply of suckler cows on rangeland producing suckled calves for finishing. The feedlots usually finish the store cattle intensively over a 6-month period, and a throughput of 150,000 animals per year is possible from a single feedlot. Such an intensity of operation raises environmental problems, similar to those caused by an excessive livestock density in parts of Holland. Local arable farmers may be contracted to produce whole-crop barley silage with some chopped hay or straw and rolled barley for the final fattening period. The cattle are kept in pens in groups of about 400, which are sloped to allow the liquids to run off into an evaporation pond. Twice a year the pens are cleared out and the manure spread on neighbouring land. Some exchange of cattle excreta for straw may be arranged with local arable farmers.

Finishing dairy cows

Dairy cows are culled for many reasons, some of which necessitate their immediate slaughter (e.g. injury, severe disease), but others allow them to be fattened before slaughter (e.g. low milk yield, failure to conceive). Care must be taken that there is no disease risk to other stock from bought-in cattle. Lame cows will not fatten well if they are unable to feed, for example at pasture. In winter cows need to be fed a small amount of concentrate, say 0.5 kg, in addition to *ad libitum* forage. Cull cows are available from both dairy and beef suckler herds, but the cows from the former will usually be more valuable as they are younger. The exception to this is cows from the Channel Island breeds, which are very difficult to get in a fat condition that is suitable for slaughter. Prices tend to be high in spring because few cows are culled at this time. Many autumn-calving cattle are slaughtered at the end of their lactation in the late summer. Thus, if cull cows are purchased in spring and take to the following autumn to get into a sufficient condition for sale, the extra weight gain may be offset by a reduced price per unit weight. Culls purchased in autumn, however, have to be kept for several months to put on weight and fetch a higher price in the spring. Feeding them is more expensive in the winter months as conserved forage is usually used, whereas in summer in temperate conditions, the feeding is cheaper as the cattle are out at pasture. In Britain, the dairy cow finishing trade has fluctuated considerably with the change in beef demand as a result of the BSE crisis, and the compulsory slaughter of older cows ended the business temporarily. It is the nature of the trade and individuals involved that other enterprises can often be substituted to utilize resources. Before the BSE crisis, a few farmers had organized themselves into cull cow-buying syndicates, which led to fewer marketing problems as the syndicate could match their supply to market requirements. In countries outside the EU, dairy culls may be implanted with hormone analogues, principally synthetic androgens, but this should only be after lactation has ended. The hormone implant can increase weight gain by about a quarter and improve the lean to fat ratio.

Indoor/outdoor rearing of beef cattle

In most countries, the climate is too harsh for cattle to continue growing adequately if they are outside during winter, so they are brought inside and fed conserved feed. Calves are usually purchased after weaning, and in the case of autumn-born, early maturing breeds, they are reared over an 18-month period. The later maturing breeds, particularly those of Continental origin such as the Charolais and Limousin, take longer to reach an adequate fat class and are usually reared for up to 24 months, which involves a second summer at pasture for autumn-born cattle. In the UK, where the dairy and beef industries are closely linked, the most common cattle used in this system are dairy × beef steers. Many dairy farmers run both a dairy unit and beef fattening system. Dairy cattle, such as the Friesian and Channel Island breeds, are different to most of the beef breeds, in that they put on more intermuscular and less subcutaneous fat. This may be desirable for cooking, but it means that if cattle are selected for slaughter on the basis of a fixed subcutaneous fat score, the Friesian or Friesian cross will have a greater total fat content. Among the beef breeds, there is no evidence for a relationship between growth rate and efficiency of production, so the main consideration is for the farmer to use a breed that suits his system.

When finishing cattle over an 18-month period, target growth rates are about 0.8 kg day^{-1}, for beef × dairy cattle. These might be reduced by about 0.1 kg day^{-1} for the first 6 months, as the young animal is not capable of growing so fast without a highly concentrated diet. If growth is less than 0.7 kg day^{-1} over the first 6 months, they will compensate to some extent when they are turned out to pasture, but the final weight will still be less than cattle that have grown rapidly throughout. Most farmers manage to achieve adequate performance in the cattle when they are housed, but growth rates when the cattle are at pasture are often disappointing. Effective pasture and grazing management is considered in more detail in Chapter 4.

High growth rates in housed cattle are best achieved by offering high-quality forage *ad libitum*. If this is not available, whatever forage is available should be supplemented with a cereal, such as rolled barley, the quantity depending on the quality of forage fed (Table 8.5). At all costs, a farmer should avoid the scenario where he plans to finish the cattle at 18 months, finds that growth rate is insufficient after he has utilized expensive feed during the winter months and decides to turn the cattle out to finish them at pasture. This will prolong the finishing period considerably, make other cattle short of pasture and make the enterprise unprofitable. Sufficient concentrate must be fed to allow the cattle to finish indoors, if this is what was planned. If insufficient concentrate is fed on a daily basis early on in the winter, the farmer may actually end up feeding more concentrates in total, because he cannot start marketing his cattle in the mid-winter period (Table 8.6). The successful operator knows how fast his cattle are growing and feeds

Table 8.5. Effect of silage quality on the requirements for a cereal supplement by cattle during the finishing winter for a growth rate of 0.8 kg day^{-1}.

	Silage organic matter digestibility ('D' value)		
	65	60	55
Silage (t)	5.5	4.4	3.4
Barley (t)	0.3	0.6	0.8

Table 8.6. An example of the potential increase in total concentrate requirement for finishing beef cattle caused by an inadequate daily concentrate allocation during the second winter period.

	Rolled barley (kg day^{-1} DM per head)		
	2	2.5	3
Silage (kg day^{-1} DM)	5.3	5.0	4.8
Live weight gain (kg day ^{1})	0.7	0.8	0.9
Finishing period (days)	300	255	165
Total silage requirements (t per animal)	6.0	4.4	3.2
Total barley requirements (t per animal)	0.67	0.62	0.53

supplements accordingly, enabling him to market his cattle at the right time and plan for the next season's cattle. Regular weighing will allow growth rates to be monitored and is an important discipline in cattle finishing.

If cattle are reared over a longer period, such as a 24-month system that finishes the autumn-born cattle off pasture in their second summer, the farmer must ensure that he does not feed too much expensive food during winter. Gross margins are usually less than for 18-month finishing as the overheads are covered for a longer period. Growth rates are usually reduced to about 0.5 kg day^{-1} during the second winter, and the cattle will then compensate when they are at pasture, growing at up to 1 kg day^{-1}. If they grow faster than this during their second winter, their performance at pasture is likely to be disappointing.

The system can be run with a mixture of late and early maturing cattle, with the early-maturing cattle being marketed in the middle of their second summer at *c.* 20 months, leaving the remaining cattle more pasture so that they can grow adequately to finish in the late summer period. A leader–follower grazing system can also be utilized for this system, with the first year cattle grazing ahead of the second year cattle, but this is not possible for an 18-month rearing system.

Indoor rearing (intensive beef production)

Intensive beef production usually refers to systems where the cattle are fed indoors throughout their life, and includes cattle fed on cereal-based diets and those fed on mainly conserved forages.

The main advantages of intensive systems are that firstly, in highly populated regions bulls can be safely housed indoors but often not outdoors, and their faster growth and high potential to put on lean meat tissue can be utilized, and secondly that the feeding period is reduced to below 1 year, allowing annual turnover of stock. Later maturing breeds of cattle, however, can be difficult to finish within a year unless a high-quality forage is fed and concentrate supplements given. With most breeds of medium- to late-maturing bulls, growth rates in excess of 1.1 kg day^{-1} should be expected. The main disadvantage of the system is the high working capital requirement, in buildings and machinery (Table 8.7). However, the working capital requirement is more evenly spread throughout the year, as only one group of cattle is on the farm at one time, compared with two groups for most less-intensive systems.

A specific advantage with silage-based systems is that the stocking rate is increased, compared with systems that involve grazing, because the grass grown is usually better utilized when conserved as silage than grazed. Losses of 20% are possible for ensiled grass (see Chapter 9), but are commonly more than 30% when grass is grazed. Grass and maize silages are most common, or a mixture of the two, since the high protein concentration in grass will complement the high energy content of maize. Roots can be fed, but not usually at more than one-third of the diet. Calves are usually initially reared on hay and transferred to a silage diet at 8–10 weeks. Protein supplements can be kept at a constant level, so that as the cattle grow they consume more silage and the protein content of the ration is reduced.

Several problems with silage beef systems may occur and a good management system is needed.

- Bulls can be difficult to handle and may engage in damaging behaviours, such as riding each other. Excessive riding can only be overcome by removing the animal that is being bullied in this way. It seems to be a form of redirected aggression and does not necessarily have a sexual function. To avoid this problem bulls should not be mixed when they are older than 6 months and should be kept in groups of less than 20. Problems with aggressive animals may be exacerbated by keeping the cattle on slats, as a high stocking density is required for the faeces to be pushed through the slats, or by hot conditions in the buildings. Slats also cause more lameness in the cattle. Specific care must be taken when taking the bulls to slaughter to avoid dark cutting.

Table 8.7. Performance and capital requirements of cattle on storage and 18-month feeding systems.

	18-month feeding	Storage feeding
Live weight gain (kg day^{-1})	0.77	1.02
Time to finishing from 12 weeks of age (days)	466	355
Concentrate requirements (kg)	809	846
Stocking rate (animals ha^{-1})	3.4	6.2
Utilized metabolizable energy (GJ ha^{-1})	67	99
Relative gross margin (18 months = 100)	100	163
Relative working capital (annual mean per head 18 months = 100)		
Annual mean per head	100	88
Annual mean ha^{-1}	300	528
Peak ha^{-1}	416	528

- Silage feeding in summer can be difficult because of spoilage in hot weather. Farmers should use a long narrow clamp for summer feeding and preferably feed out with a silage block cutter that will leave a tidy face. If the silage is teased out of the clamp, air enters and secondary fermentation occurs. Pits may need to be open at both ends, so that one end can be filled while silage is being fed out from the other end. Good silage-making machinery needs to be used so that the quality is high. With so much reliance on silage quality, the best techniques need to be used and an additive to speed the fermentation is advisable.

The system of feeding a predominantly cereal diet is not common, except where the two main inputs, cereals and calves, are inexpensive relative to the finished product. More efficient meat production is obtainable by feeding cereals to poultry or pigs and there is less risk of digestive disturbances. However, some farmers keep cattle in this way, particularly in hot climates where forage may expensive and of low quality. The main advantage is the high throughput and rapid fattening of bulls that is possible. Heifers do not put on sufficient weight on this system because they become fat too early. The main metabolic disorders that occur are as follows:

- rumen acidosis, caused by rapid degradation of cereals by bacteria;
- bloat, particularly on ground rations and when the cattle overeat, often caused by offering the feed in limited quantities twice a day rather than *ad libitum*;
- liver abscesses, caused by damage to the rumen wall which allows bacteria to enter the blood stream. In chronic cases this causes liver damage and condemnation in the abattoir, in acute cases it may cause death;
- laminitis, caused again by excessive acidity in the rumen, leading to separation of the laminae of the hoof, haemorrhaging in the heel bulb and a severe and painful lameness.

Heifer rearing

Heifer calves born on a dairy farm are mostly reared for potential cow replacements rather than for meat. The extent to which heifers are used for replacements depends to a large extent on the culling rate of the cows, but on an average farm in the UK, 25% of the cows will be culled annually and must be replaced by a heifer, or young cow. In some countries it is up to one-third, i.e. a cow only lasts three lactations on average in the herd, which is because of the high expectations in terms of productivity, ability of the cow to resist disease and to rebreed on an annual basis. The point may have been reached where increases in the intensification of cow production are counterproductive as culling rates are increased, a problem being addressed by the cattle breeders by including lifetime production, rather than yield in one lactation, into the breeding indices (see Chapter 5).

The length of the rearing period also determines the number of replace-ment heifers that are required annually. On many farms this has been intensified to 2 years, whereas it often used to be 3 years. Cows calving for the first time at 3 years of age give more milk in their first lactation but not on a lifetime basis. Again any further intensification tends to be counterproductive. Israeli scientists have found that Friesian heifers could quite satisfactorily calve at 18 months, which would be beneficial in their intensive production systems. However, the cows were too small to join the other cows in the milking herd, and the large amount of growth required during lactation meant that milk yields were disappointing. The length of the rearing period must depend on the farmer's resources. If he has good quality grazing and feeding systems for winter-feeding, first calving at 2 years of age is almost certainly best. However, if only poor-quality grazing is available for the heifers, first calving at 3 years is usually best. It runs the risk of the heifers being too fat, which could cause calving difficulty, but so can calving heifers at 2 years when their pelvis may be too small for the calf to pass through. Calving at 3 years of age also requires 75% more land than 2-year calving and 30–40% more working capital, with higher fixed costs and interest payments.

It is not essential that dairy farmers rear their own replacements. If they buy in heifers that are close to calving, they can upgrade the average milk yield of their herd if the heifers are of good genetic stock, but they run the risk of buying diseased cattle. The farmer is however, able to ensure that well grown animals are purchased that have achieved target weights on time. This policy may be favoured because it releases buildings for other operations. However, most farmers like the security of rearing their own livestock and they take a pride in the quality of the replacement cattle reared. Usually, only farmers that do not have adequate facilities buy in replacement stock.

In the first few weeks of life, heifers that are being reared for replacements follow a similar system to beef cattle. At birth, the herdsperson should ensure that the calf breathes normally and that it suckles. The navel should be dressed with an antiseptic when the calf is born indoors, but some farmers like to leave it to dry up naturally when the calf is born outdoors and it is dry and sunny. After 1–3 weeks the calf should be disbudded (dehorned) if it is not of a polled breed. By 4 weeks of age, any surplus teats should have been removed and the calf identified by means of an ear tag or other mark, such as a brand. Target growth rates are less than for steers or bulls, 0.6–0.7 kg day^{-1} for the first 6 months if they are indoors, 0.6 kg day^{-1} for their first summer, 0.5 kg day^{-1} for their second winter and finally 0.7 kg day^{-1} for their second summer at pasture. The success of the first summer growing period depends largely on adequate prevention of stomach worm infestation. However, whereas the aim with male calves is usually to prevent the animal being exposed to gastro-intestinal worms, heifers have to build up immunity by the time they are adult, so gradual, careful exposure is necessary. At 15 months, the heifers will need to be served if they are to calve at 2 years of age. There is no difficulty in getting them to conceive at a light weight – they reach puberty at only 45% of their mature body weight, i.e. about 10–12 months for a Friesian. However, a

Friesian will need to be at least 350 kg for it to be large enough at calving – 520 kg pre calving, 470 kg post-calving. For the large Friesian Holstein cows that are favoured today, these weights need to be increased by up to 20%.

When close to calving, the heifers need to be accustomed to entering the milking parlour and, if appropriate, being fed there. They should have been taught to lie in cubicles in their early years, and before calving, they should be gradually introduced to concentrates to accustom their rumen flora to the new feed.

Conclusions

Cattle rearing requires the skill and dedication of a good stockperson. The rewards are considerable, in the satisfaction of producing an animal that will produce valuable meat or enter a dairy herd to produce milk for most of the rest of her life. The demands of the general public for high-quality produce and a high quality of life for the cattle make excellent stockmanship essential. At the same time, the need for efficiency has never been greater, with an ever-growing world population to feed, the creation of a good foundation stock for milk production and efficient utilization of land resources that would otherwise be wasted is an essential contribution to the world food supply.

References

Earle, D.F. (1976) A guide to scoring dairy cow condition. *Journal of Agriculture, Victoria* 74, 228–231.

Edmondson, A.J., Lear, I.J., Weaver L.D., Farver, J. and Webster, G. (1989) A body condition score for Holstein dairy cows. *Journal of Dairy Science* 72, 68–78.

Lowman, B.G., Scott, N. and Sommerville, S. (1976) Condition scoring of dairy cattle. *Bulletin of the East of Scotland College of Agriculture*, No. 6.

Scott, J.D.J., Lamont, N., Smeaton, D.C. and Hudson, S.J. (eds) (1980) *Sheep and Cattle Nutrition*. Ministry of Agriculture and Fisheries, Agricultural Research Division, Wellington, New Zealand.

Further Reading

Allen, D. (1990) *Planned Beef Production and Marketing*. BSP Professional, Oxford.

Allen, D. (1992) *Rationing Beef Cattle: a Practical Manual*. Chalcombe Publications, Canterbury.

Bent, M. (ed.) (1993) *Livestock Productivity Enhancers: an Economic Assessment*. CAB International, Wallingford.

Lawrence, T.L.J. and Fowler, F.R. (1997) *Growth of Farm Animals*. CAB International, Wallingford.

Preston, T.R. and Willis, M.B. (1975) *Intensive Beef Production*. Pergamon Press, Oxford.

Thickett, W., Mitchell, D. and Hallows, B. (1995) *Calf Rearing*. Farming Press, Ipswich.

Cattle Production and the Environment

9

Introduction

The cattle industry has often been the subject of criticism with respect to its impact on the environment. In South America, the destruction of large areas of rain forest to create grassland for cattle grazing is held partly responsible for global warming. In North America and parts of Europe, the imbalance between waste production by the animals and the availability of land on which to spread the waste is believed to contribute to pollution of water supplies. In parts of Africa, cattle contribute to overgrazing and the treading and removal of plant cover in hill regions causes soil erosion. Even the typical British family farm of 50 dairy cows has a potential pollution load equivalent to that from a human population of 500 people.

At the same time, cattle are acknowledged to perform a useful function in effectively making fibrous grasses into food for human consumption in areas where crops for direct human consumption cannot be grown. They also produce valuable manure to fertilize the land or to be burnt as fuel, saving the trees from destruction for firewood. In desert reclamation programmes, the installation of cattle farms may be the first action to be taken as their manure will stabilize the sandy soil and increase water retention capacity. In the grazing situation, cattle are often preferable to sheep or goats on marginal land, as they do less destruction and cannot graze as close to the ground, thereby leaving a greater plant cover. They are less selective in their grazing habits because of their broad muzzle, so that they cannot selectively consume species that could become depleted in the sward. They are a major source of traction in developing countries, reducing the reliance on tractor power and hence fossil fuel use.

A new emphasis on sustainable agricultural systems is being created in many regions of the world, which is ensuring that the systems of cattle production practised are those that allow the benefits to outweigh the disadvantages, particularly by new entrants to the industry. Some governments are helping these changes, with assistance for farmers that wish to practice cattle production in ways that are not as profitable as intensive farming but are more beneficial for the environment. The assistance for organic farmers in Europe is one example of this. Although in some regions there would not be enough land for all people to eat organic cattle products, it is still justifiable for farmers to receive a subsidy for managing it in a manner that both improves the environment and produces cattle products in a safe and sustainable way.

Controlling Carbon, Nitrogen and Phosphorus Emissions

Intensification of the cattle production industry has only been possible with large inputs of fossil reserves, principally fertilizers and fuel, which allow food production from the land to be increased and a larger number of cattle to be kept on a farm. In addition, considerable quantities of concentrates are purchased from arable farms, which further intensifies the production from the livestock area. This intensification of cattle production onto small areas of land, while being generally advantageous in terms of labour use and other economies of scale, may produce problems with waste disposal. The ability of the disposal sink, such as the soil or the ground water, to detoxify and utilize the wastes is easily overloaded and emissions may escape into the public water supply or the atmosphere. The problems do not end when the cattle products leave the farm. Inputs of fossil energy, relative to home-grown energy, are particularly high once the animal has left the farm.

Both carbon and nitrogen compounds are important greenhouse gases, principally carbon dioxide, methane, nitrous oxides and ammonia. However, the control of emissions from the cattle farm is usually focused on nitrogen products. The control of methane emissions is possible, since feeding a more concentrated diet reduces methanogenesis in the rumen and promotes propio-genesis. However, the increase in the use of cereals and other high-energy feeds required to be imported onto the farm could increase fossil fuel use and consequently reduce the health of the cattle if inadequate fibre is consumed. Similar reductions in methanogenesis could be achieved by adding iono-phores to the diet, but there are concerns over residues, especially in milk.

Nitrogen utilization efficiency on dairy farms is typically 15–35% (Kristensen and Halberg, 1997) and has decreased in recent years with the increase in use of nitrogen fertilizer, for example by one-third over the last 30 years in Holland. However, the calculation of utilization efficiencies can be misleading, at least in short-term calculations, since the emissions are often lost not from the transient pool of a nutrient, but from the very substantial reservoir of the nutrient in the source. For example, the flux of nitrogen into leached water may be 300–400 kg ha^{-1} in an intensive dairy farm, i.e. 80% of the amount of nitrogen applied. However, some of the release of nitrates into ground water comes from the soil organically bound nitrogen pool, which typically contains 7000 kg N ha^{-1}. The best dairy farms can have nitrogen surpluses as low as 75 kg N ha^{-1}, but even on low intensity farms accumulation is typically about 225 kg ha^{-1} (Fig. 9.1; Kristensen and Halberg, 1997).

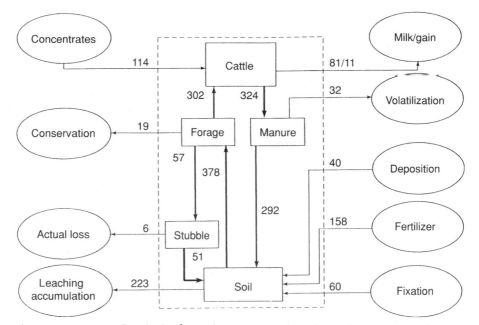

Fig. 9.1. Nitrogen flux (kg ha^{-1}) in a low-intensity dairy farm, showing inputs from concentrates, fertilizer, aerial deposition and fixation by soil microbes and losses through volatilization, conservation processes, and from the herbage stubble to the soil and the air. The remainder is accounted for by output in the form of milk production or live weight gain and leaching or accumulation in the soil reservoir (from van Bruchem *et al.*, 1997).

Some nitrate leaching is probably inevitable, but good practices can still be adopted which will help to minimize leaching. This mainly concerns the timing of nitrogen applications to avoid the periods of heavy rainfall and low plant growth, when nitrogen uptake is reduced. Following the application of nitrogen fertilizer, the first source of nitrogen loss occurs from nutrient runoff from the surface of the field, into watercourses. This is most likely after heavy applications or when the soil is waterlogged or frozen. Sloping ground will increase the risk of nitrogen runoff. Nitrogen can also be volatilized to ammonia before it is absorbed into the soil. Once in the soil, the nitrogen is prone to leaching losses, which is most likely when the rainfall is high. If the nutrients are leached below the rooting zone of the grass plants, they can never be absorbed by the plant. Typically about three-quarters of the nitrogen applied as slurry in autumn will be lost by leaching, runoff and to the atmosphere, for a winter application this would be reduced to one-half, and for a spring application one-quarter. However, to store the production of slurry for the winter months from a dairy herd, a large store is required. Over-flowing slurry stores are a significant cause of watercourse pollution.

Ploughing is particularly likely to increase nitrate leaching, so permanent grassland is favoured in preference to temporary grass leys or arable crops.

Maintaining a crop cover for as much of the year as possible is important, especially in countries such as Britain where the rainfall is not strongly seasonal and is unpredictable.

The cattle industry in Holland represents one of the most extreme examples of the pollution problems caused by intensive livestock farms. It is concentrated in the north of the country, so that the transport of concentrate feeds from the ports is reduced and it can occupy land that cannot be used for other purposes. However, the stocking density is so high that nutrient emissions are too large for the land to absorb safely. As a result, the Dutch must now devise alternative production systems that reduce emissions, in the case of nitrogen and phosphorus, to below 180 and 8.7 kg ha^{-1} year^{-1} respectively. An intensive dairy system may have losses of 400–500 kg N ha^{-1} year^{-1}, but this can be reduced to about 200–250 kg N ha^{-1} year^{-1} by reducing inputs of N fertilizer and adopting environmentally friendly practices. Such possibilities are exemplified by a comparison of an intensive dairy farm in Holland with a mountain dairy farm in Italy producing milk for Parmigiano–Reggiano cheese under local regulations (Table 9.1). These regulations prohibit the feeding of silage, industrial by-products and a range of feed ingredients, because of adverse effects on cheese quality. Raw milk is collected from the farm twice a day and is processed under strict conditions. The nitrogen balances show that, although the proportion of useful output

Table 9.1. Nitrogen balances of an intensive dairy farm in the Netherlands and a mountain farm in Italy producing milk for the production of specialist cheese under local regulations that control the farming methods (from de Roest, 1997).

	Netherlands intensive dairy farm		Italian specialist dairy farm	
	kg ha^{-1}	%	kg ha^{-1}	%
Inputs				
Cattle purchases	4	0.7	1	0.2
Straw	1	0.2	7	2.2
Fertilizers	346	62	36	11
Organic manure purchased	3	0.5	22	7
Nitrogen deposition	42	8	18	5
Nitrogen fixation	4	0.7	22	7
Roughage purchased	45	8	85	26
Concentrates purchased	114	20	132	41
Milk powder	1	0.2	1	0.4
Total inputs	560	100	324	100
Outputs				
Cattle	16	19	8	10
Milk	64	76	54	65
Manure	3	4	9	10
Others	1	1	13	15
Total outputs	84	100	84	100
Balance	476		240	

(cattle, milk, manure) as milk is greater in the Dutch farm, the balance or accumulation of nitrogen on each hectare of the farm is nearly double that of the Italian mountain farm. Mountain farms have a greater input of labour per cow, but much of this is family labour. If this is costed at normal rates, the production cost of the milk output is high, but this may be compensated by the greater product value, as can be seen in a comparison of extensive mountain and intensive lowland dairy production in Italy (Table 9.2). Hence, the mountain farm provides employment in a marginal economic region, as well as preserving the environment for future generations. However, the mountain farms have to rely more on purchased concentrates than lowland farms, which can grow forages more easily. Hence, for the production of high quality foods, the type of concentrate used must be strictly regulated.

Forage production on a reduced input farm can be maintained by making better use of cattle excreta and perhaps mixing it with straw to make manure before spreading on the land. Farmyard manure has slow nitrogen release characteristics but also contains useful amounts of phosphorus, calcium, magnesium, sulphur and trace elements, all differing from the supply in fertilizers by their long period of availability in the soil. Urine is particularly rich in nitrogen and potassium. The use of legumes is encouraged, but forage crude protein contents of more than $180 \, g \, kg^{-1}$ dry matter (DM) will not enable the farm to be below the emission limits, as well as potentially reducing reproductive rates of the cows. Improving the efficiency of nitrogen utilization by cattle, for example by matching the energy supply to the protein breakdown, will have some impact, but not as much as reducing nitrogen inputs to the farm.

Phosphorus emissions are even more difficult to control and need to be tackled by managing farmyard manure properly, minimizing phosphorus fertilizer use and reducing purchased concentrate use. The main problem is surface runoff from farmyard manures (Fig. 9.2), which ends up in

Table 9.2. A comparison of the technical and economic efficiency of mountain dairy farms for the production of Parmigiano–Reggiano cheese and intensive lowland dairy farms producing milk for liquid consumption in northern Italy (from de Roest, 1997).

	Mountain dairy farm	Intensive lowland dairy farm
Number of cows	25	70
Cultivated area (ha)	27	35
Cows per ha forage crops	1.1	2.5
Milk yield per cow (kg year^{-1})	4800	6200
Concentrates per cow (kg year^{-1})	2100	1700
Working units (people year^{-1})	2.4	3.5
Costs per kg milk (lire)		
Concentrates	212	156
Family labour	390	120
Hired labour	1	40
Total production cost	910	550

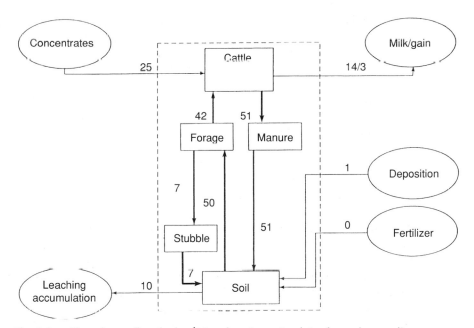

Fig. 9.2. Phosphorus flux (kg ha^{-1}) in a low-intensity dairy farm, that applies no phosphorus fertilizer, showing input from concentrate and aerial deposition, and output as milk production or live weight gain, or as leaching or accumulation in the soil reservoir (from van Bruchem *et al.*, 1997).

watercourses and causes eutrophication in lakes. Eutrophication is the term used to describe the depletion of oxygen reserves in the upper warm water regions of a lake (the epilimnion) as a result of excessive plant growth and organic matter decay. This is most probably caused by phosphorus runoff from manures spread on the land or stored near a watercourse, but it can also be caused by nitrogen deposition from volatilized ammonia. The limiting nutrient in seawater is more likely to be nitrogen than phosphorus. Phosphorus fertilizers also have to be carefully controlled because of their high cadmium content. The cadmium content varies considerably, but may be as high as 150 mg Cd kg^{-1} P, in which case regular application could increase herbage cadmium content above the EU legal limit of 1 mg kg^{-1} herbage DM.

The application of fertilizers can be made more efficient by applying the optimum compounds at the correct rates to each area of land, taking into account the soil type, crop type and weather. This requires detailed and up-to-date soil maps for each field on the farm, precision application and a knowledge of past and forecast weather patterns. The benefit of such high-technology inputs into fertilizer application is that growth can be optimized with the minimum of inputs. It will be more difficult for mixed crops, such as grass/clover mixtures, where the requirements of the species are different at the various stages of the growing season. Fertilizers that are mixed for optimum growth of the crop at each stage in its production cycle are likely to contain more than just nitrogen, phosphorus and potassium – the three nutrients most commonly applied. Sulphur may be co-limiting with nitrogen and sodium, but sulphur applications should be restricted as high concentrations in herbage reduce palatability and milk fat concentrations. Although sodium does not greatly enhance grass growth in most temperate conditions, it will increase the palatability of the grass and cattle intake. Its use can replace some potassium fertilizer, which is required more by the plant than the animal, with the benefit that the animal's needs for sodium are more effectively met.

An efficient fertilizing strategy should aim to reduce fertilizer application, but to tailor specific fertilizers to the requirements of each field. It is doubtful whether fertilizer application can ever be eliminated in the long term. As long as crops are removed from the land, there will be a net drain of minerals from the system, and many agricultural systems have failed in the past because the land becomes exhausted and nutrient-deficient. The most sustainable low fertilizer-input system is one with grazing cattle, since they return many of the nutrients that they consume. Although the nutrient release from faeces is slow, the urine contributes to point source losses of nitrogen as it is deposited in small concentrated areas, contributing to leaching of nitrogen at these points. However, in terms of environmental risk, the nitrate leaching from permanent grassland is not a major cause for concern, since it is less than from ploughed fields. The major concerns with excessive nitrogen fertilizer use in grassland systems, relative to the nitrogen output, is the loss of nitrogen to the atmosphere through denitrification and the fossil fuel use during the fertilizer manufacturing process.

Slurry Treatment

A dairy cow produces approximately 60 l of faeces and urine per day, or 0.06 m^3. This is usually collected into a semi-solid mixture, or slurry, containing excreta to which waste water is added, for example from washing cattle yards and stored in a tank. Some dilution is necessary for efficient storage, handling and spreading on the land. A winter rainfall of 500 mm on a 0.5 ha farmstead will produce 2500 m^3 to be stored, i.e. a volume roughly equivalent to the slurry from a 320-cow dairy herd. This will produce a suitable degree of dilution for handling purposes.

Slurry will flow under gravity, a physical characteristic that can be used to collect it into a central pit and minimize manual or mechanized movement of the substance. Most dairy farms with loose housing of their cows now produce slurry rather than farmyard manure, because the former is more easily handled mechanically. Slurry is scraped out of cubicle passageways either by a tractor with a rubber blade mounted on the back or by automatic scrapers that pass down the passageway approximately every hour attached by a chain. Scraping with a tractor should normally be done twice a day, around milking time, otherwise there is too much slurry accumulated in the passages. After scraping it out of the building, the slurry is scraped by tractor to a pit, from where it may be pumped to an above ground store. More effective storage of cattle excreta is required on many farms. The older types of slurry stores had gaps in between the wall panels, creating a 'weeping' wall from which the more liquid component of the slurry could emerge and disappear into the soil. Nowadays, storage tanks should have sealed walls and be covered to prevent rainwater entering. It may need to be stirred to reduce crust formation and stimulate aerobic fermentation, which will reduce the odour and the biochemical oxygen demand (BOD; Table 9.3). However, if the store is not covered, stirring will increase loss of nitrogen in the form of ammonia and nitrogen gas after denitrification. Typically the nitrogen content of slurry is reduced by one-third during storage. The liberated ammonia will enter the atmosphere and when it eventually is returned to the land it will act as a fertilizer. This is of most concern in hill areas where tree growth is stimulated and the trees become more susceptible to disease; also, nitrification of the ammonium in poorly buffered soils will cause acidification of the soil. Apart from sealing the store, gaseous losses can be prevented by adding nitrification inhibitors. Nitrification is responsible for nitrogen losses from the soil, where NH_4^+ ions, which are not readily leached as they are adsorbed to clay particles, are converted into nitrates, which are readily leached. The most common nitrification inhibitor is dicyandiamide, which acts for between 2 and 6 months to prevent nitrification. When added, it has produced reductions in nitrogen losses in countries with a cold winter, but there is some doubt if it would survive a mild British winter.

If slurry is spread near to watercourses and it has a high BOD, it will deplete the water's oxygen content, making it difficult for fish and other aquatic organisms to survive. Most old dairy farms are sited near to water sources, often a spring, so that water was available for the farming operations and the farmer's household. The risk of polluting the water supply in these circumstances is considerable. Runoff control can be achieved by constructing a drainage ditch around the farm, which diverts runoff into a holding pond. Periodically, and especially after a period of high rainfall, the water from the holding pond should be spread onto the land. The pond should have the capacity to hold a rainfall incident equivalent to the largest incident experienced over the last 10 years.

Table 9.3. The biochemical oxygen demand (BOD) of substances produced on cattle farms.

Substance	BOD (mg l^{-1})
Dirty water (dairy parlour and yard washings)	1,000–2,000
Liquid wastes draining from slurry stores	1,000–12,000
Cattle slurry	10,000–20,000
Silage effluent	30,000–80,000
Milk	140,000

Waste disposal must be paramount in choosing a site for a cattle farm nowadays. It involves investigating the soil type, local climate, surrounding crops and proximity of human population. Sandy soils are more susceptible to leaching losses than clay soils or loams. If there are no adequate days when the ground is frozen to allow slurry tankers to spread on to the land, a large store will be needed to hold the slurry produced during the winter. The length of the grazing season also needs to be assessed so that winter storage can be determined. Different crops have different requirements for nitrogen, and so can absorb different slurry applications. For permanent grassland, a slurry application in spring can cause capping on the grass and loss of sward-production potential. Applications to any crop in the autumn should be avoided because of the high leaching risk. The high potassium content of slurry can also be a risk on grassland as potassium in herbage inhibits the absorption of magnesium, potentially causing hypomagnesaemia in cows' grazing lush pasture.

One of the major problems of spreading slurry near human habitations is the noxious odour. This is often exacerbated by the use of slurry tankers fitted with a discharge nozzle delivering to a splash-plate that spreads the slurry in an arc behind the tanker. This creates a small droplet size that increases both odour release and volatilization of compounds into the atmosphere. Pathogenic microbes, such as *M. bovis*, may be spread several hundred metres and could potentially infect humans or livestock. Slurry injectors normally deliver the slurry directly into the soil at a depth of 150 mm, via a series of hollow tines fitted with wings to aid dispersal of the slurry beneath the ground. After each tine has created the injection slot, and the slurry has been injected, a wheel or roller passes behind it to close it. The reduction in odour is of the order of 70%, but grass yield may also be reduced because of the damage to the sward. However, the tractor power requirements are significant, leading to carbon dioxide emissions from fuel use. Slurry injection is not possible in stony soil or in hilly terrain. Shallow injection is also possible, to a depth of 60 mm, which is more suitable for grassland, and gives an adequate reduction in odours of about 50%. It has much lower tractor power requirements. In Holland, injection is the only permitted method of slurry disposal on farmland and is facilitated by the soil type in reclaimed land.

Anaerobic digestion of slurry

Slurry can be effectively digested anaerobically by bacteria to produce methane gas, an odourless liquid and friable solid. The gas can be used for cooking, although it not as pure as natural gas, the liquid can be pumped onto the land via an umbilical cord and the solid material put into bags and sold as garden compost. Digestion produces a 40% reduction in the chemical oxygen demand, which is not enough to allow it to enter watercourses, but the main advantage is that the liquid waste product can be easily applied via a pipeline onto the surrounding farmland. The pipeline can be connected to a tractor or an irrigation system. The main difficulty from the cattle farmer's point of view is to keep the digestion process at a constant temperature for the bacteria to grow, which means protecting it from variation in environmental temperature. Digestion systems are therefore most popular in hot countries where natural fuels are expensive, in China for example. In Britain, the digestion chamber needs to be heated, perhaps with the gas from the bacterial fermentation. It can be difficult to have a continuous system from which the solid residue can be extracted, and continuous flow systems that use sealed polythene chambers set in the ground often have a short life. Some governments have subsidized the installation of anaerobic fermentation plants on the grounds that they reduce emissions.

Slurry separation

An alternative treatment method for slurry is separation, which produces a friable solid material for sale as compost or fertilizer and a liquid product for spreading on the land. Separation is achieved with varying degrees of efficiency by vibrating or rotary screens, or presses using belts or rollers.

Sewage Sludge

In Britain and other countries some difficulties are likely in the future as some sewage sludge has to be disposed of on farmland, rather than dumped at sea, posing problems of nitrogen overload, public nuisance, pathogen transmission and soil contamination with heavy metals. Sludge nitrogen, being in the ammoniacal or organically bound form is not leached as readily as fertilizer nitrogen, but still in many areas there are insufficient farm sites that are suitable to take the sludge, forcing some of it to be combusted. The soil contamination with heavy metals from sludge, particularly zinc, copper, lead and cadmium, is becoming less of a problem in many developed countries as industry reduces its emissions of toxic metals into industrial effluent. Pathogens should be minimized by chemical, biological or heat treatment, but the disease risk transfer is to some extent safeguarded by controls on application of sludge to certain crops, e.g. fruit and vegetable crops normally eaten raw. The same controls may not always be in place or enforced outside the EU.

Silage Effluent

More silage is being produced for cattle than dried forages in the UK, mainly because advanced technology is now available that will make and feed silage automatically to cattle, and since it is of higher energy value than dried forages the cattle produce more milk or grow faster. The increased use of nitrogen fertilizer on grass produces a leafier, wetter crop, and the tendency to minimize the wilting period for silages for rapid conservation and low field losses reduces crop losses but increases effluent production (Table 9.4).

With the increased use of silage in the UK, there have been more pollution incidents associated with silage effluent, which now account for one-third of all UK pollution incidents. As well as government penalties, there are often prosecutions by angling associations for the damage to fish stocks. The threat to watercourses from silage effluent is actually greater than that from slurry, even though more slurry is produced, because of the high BOD of the effluent (Table 9.3). Silage effluent is often allowed to seep from silage clamps unchecked and may end up in watercourses, where it is a high risk for eutrophication of the waters. Recently the collection of the effluent from silage clamps has become more generally accepted and in some countries is legally required. The volume of effluent production (l t^{-1} herbage) can be calculated from the herbage dry matter (DM) content (g DM kg^{-1} freshweight) as follows:

$$\text{Volume of effluent} = 800 - 5 \times \text{herbage DM content} + 0.009 \text{ herbage DM content}^2.$$

Most of the effluent is produced in the first 10 days after ensiling, so the effluent tank must have sufficient capacity for this volume, as well as any rainwater that falls on uncovered clamps. A 1000 t clamp will need a tank of at least 25 m^3, and larger if very wet silage is conserved. The effluent is acidic (normally pH 4) and will etch the concrete of the clamp. Following collection, it can be either spread on the land as a fertilizer or fed to cattle or other stock. However, spreading effluent on the land may scorch crops because of its acidity. A maximum of 10 m^3 ha^{-1} should be spread, or 25 m^3 ha^{-1} if it is diluted with water at 1:1, and applications should not be repeated within 3 weeks as the soil microflora will not have had time to break it down. The fertilizer value of effluent is similar to that of farmyard manure. The crude protein content is about 250–350 g kg^{-1} DM, most of which is amino acid

Table 9.4. Typical losses (% DM) from grass silage that is either wilted in the field for 36 h or ensiled directly, both under conditions of good management (from Wilkinson, 1981).

	Direct cut	Wilted
In field		
Respiration	0	2
Mechanical loss	1	4
During storage		
Respiration	0	1
Fermentation	5	5
Effluent	6	0
Surface waste	4	6
During removal from clamp	3	3
Total	19	21

nitrogen, making it a suitable feed for pigs. The dry matter content varies from 40 to 100 g kg^{-1}, with a mean of 60 g kg^{-1}.

When feeding effluent to cattle it should be preserved by adding formalin, at 3 l t^{-1}, or acids if its pH is greater than 4. Its feeding value is equivalent to 1/20th that of barley on a fresh matter basis. It is rich in minerals, particularly potassium, and also contains ethanol. Antibacterial preservatives should be used cautiously as they may inhibit rumen fermentation if too much is consumed. There is no problem with palatability, unless the effluent has been allowed to spoil, but it is wise to offer the cattle an alternative water source. The greatest difficulty lies in the rapid production of the effluent and the cost of storing and feeding it.

An alternative to collecting the effluent as it is produced by the clamped crop, is to reduce effluent from the crop by adding absorbent material at the time of ensiling. The absorbent material can be of lower nutritional value than the ensiled herbage, such as when chopped straw is added. Straw bales can be laid at the bottom of the clamp, but are not very effective in absorbing the effluent. If straw is added, the feeding value of the final product will be reduced and also more variable, with some cattle rejecting effluent-soaked straw if *ad libitum* silage is available. Alternatively, cereal grains can be added, which will increase the quality of the finished product. These are not particularly absorbent, and the starch in the grains will not assist the fermentation of the grass, as most bacteria cannot use it as a substrate. The absorbency of the grains can be increased by grinding them and, if they can be added evenly as the grass is ensiled, a complete diet can effectively be made in the clamp.

A third possibility is to add shredded beet pulp to the ensiled crop. This is highly absorbent and will reduce the effluent production by one-half when added at about 50 kg t^{-1}. Although absorbents are effective in reducing effluent production, they are difficult to apply and may be lost in the feeding process.

Environmental Toxins and Antinutritive Agents

Cattle may be either directly affected by environmental toxins or they may be carriers of toxins in the human food chain. Some toxins also have antinutritive properties, i.e. they reduce the nutritional value of the diet but are not toxic. A description is given of some of the most significant toxic agents, although knowledge in this field is rapidly advancing and other, more potent toxins may yet be discovered.

Lead

Lead is ubiquitous in the manmade environment, because of its numerous uses. When consumed by cattle, it is toxic at relatively low concentrations compared with other livestock, about 2 mg kg^{-1} live weight, or approximately 60–100 mg kg^{-1} feed DM. It is still the most common form of poisoning of any farm livestock, with about 200 cases in cattle annually in the UK, most arising as a result of accidental consumption. Lead was, until recently, used in paints, and housed cattle often become poisoned when they lick paint in their stall. In developing countries, cattle often graze close to the road and lead, added to petrol to prevent the engine knocking, will spread up to about 10 m from the road. In arid regions, this lead is not washed off the leaves quickly and the removal of lead from petrol, now nearly complete in developed countries, has proceeded quite slowly in developing countries. The lead has toxic effects on the rumen bacteria and cattle learn to avoid lead-contaminated herbage to some extent. Once the lead has been washed off the plant leaves, it remains in the topsoil and still may be consumed, since grazing cattle commonly have up to 10% soil in their diet. The uptake by plants is low, with some entering the roots, but very little reaching the plant parts above ground. However, little of the lead will leach from the soil, presenting an almost permanent threat to grazing cattle, unless the topsoil is removed. In lead mining regions, the heaps of spoil present a threat to cattle, as lead contents remain dangerously high and will do so almost indefinitely unless remedial action is taken. Removal of the topsoil is the best way to allow such areas to be safely grazed. The areas surrounding old munitions works present a similar threat, which may only surface during a drought when cattle consume a significant quantity of soil because of the herbage being short. Discarded car batteries in a field may be licked by cattle or may even make their way into a complete diet if picked up by a forage harvester. Clay pigeon shooting may also leave lead on fields, which cattle can ingest during dry weather.

Lead has a particular affinity for bones and causes osteoporosis; it also enters the liver and kidney. It interferes with iron metabolism and may cause anaemia. Often it causes a typical blue line at the junction of the gums and teeth, and grey faeces. Most of the symptoms relate to the neurotoxicity of lead. Affected cattle may charge around and press their heads against a wall, and later they develop ataxia.

Fluorine

Fluorine is involved in bone and teeth formation. There are some areas of the world where fluorine concentrations are naturally high in deep well waters, but most fluorine toxicity arises from exposure to emissions from the processing of rock phosphates high in fluorine. Aluminium, bricks, tiles, steel and rock phosphate quarries can all produce high fluorine emissions, and the degree of exposure will depend on the prevailing winds and height of the emission source. The inclusion of phosphates in mineral supplements will add significantly to fluorine intake unless defluorinated phosphates are used.

Cattle are the most susceptible of the farm livestock to fluorine toxicity, and especially dairy heifers as their bones and teeth are actively growing.

Mottled and malformed teeth and misshapen bones are the usual symptoms of fluorine toxicity, if concentrations in the feed reach 30–40 mg F kg^{-1} feed DM. Cattle become lame, milk production can be reduced and fertility impaired at high exposure levels (>50 mg F kg^{-1} feed DM). Accumulation in bone tissue provides cattle with some protection from the toxic effects, however, once bone fluorine content reaches 30–40 times its normal level, the excess fluorine invades the soft tissues. The kidneys can excrete a certain amount, but once this is exceeded a severe anorexia ensues and death may follow.

Cadmium

Cadmium is a cause for concern both at point sources, particularly around metal smelting works, and because of the gradual accumulation in many pastures. It is deposited on the land mainly from phosphate fertilizers and sewage sludge and may also be consumed in mineral supplements with high phosphorus contents. The cadmium content of some soils is naturally high, but most of the potential problems are manmade. The problems lie not so much in its toxicity to cattle as to humans consuming the kidneys and, to a lesser extent, the livers of cattle that have built up a cadmium load over their life. Only a very small part of the ingested cadmium is absorbed, this being dependent largely on the animal's zinc status. Following absorption, cadmium is complexed with metallothioneins in the liver and gradually released to the kidney, where it is liberated by the lysosome system. It is this liberated cadmium which can cause damage to the proximal tubules. The long half-life of cadmium means that this is normally only a problem with older animals. Ingested cadmium does not readily transfer to cow's milk, so the main problem concerns cadmium in the human food chain that derives from the kidney and livers of cattle.

Dioxins

This term is commonly used for polychlorinated dibenzo-para-dioxins, dibenzofurans and polychlorinated biphenyls, although it should strictly be reserved for the compound 2,3,7,8-tetrachloridibenzo-para-dioxin. They are used in industrial chlorination processes, incineration of municipal wastes and herbicide production. There is a significant concentration in sewage sludge, which will increasingly be used on the land or burnt, to replace disposal at sea. Both of these methods of disposal could contaminate cattle products, although the concentration is forecast to decline in future years in the UK. The health risks are principally their carcinogenic, immunomodulatory and teratogenic properties that have been demonstrated in rodents, but not yet conclusively in humans. Concern arises in humans consuming cattle products, such as milk, after cattle have absorbed the chemicals directly or indirectly. This may occur in, for example, cattle lying on or eating newspapers, but surface contamination of herbage in industrial zones can cause indirect contamination. Milk products are particularly implicated because of the lipophilic nature of dioxins and the high fat content of most milk products.

Mycotoxins

Mycotoxins are sometimes present in purchased milk, but as with other contaminants they have to survive the processing and the animal's detoxifying mechanism. The main mycotoxin capable of entering milk is aflatoxin B1, which is often present in cattle feed grains. Aflatoxins may be hepatotoxic, mutagenic, immunosuppressive and carcinogenic, and there is an increase in bacterial infections of cows consuming aflatoxin. Zeolites may be used to reduce the toxicity by tight binding of the aflatoxins in the gastrointestinal tract. In Europe, the legal limit of aflatoxin B1 is 10 μg kg^{-1}, however, this may result in more than 10 ng kg^{-1} of aflatoxin M1 in the milk, the legal limit in infant formula. To ensure that the legal limit of infant milk is not exceeded, the concentration of aflatoxin B1 in cattle food should not exceed 2 μg kg^{-1}.

Conclusions

The relationship between cattle production systems and the environment should be of primary consideration in all units, and particularly when new units are being planned. Projected budgets should take into account a greater control of pollution from cattle units and the necessity that emissions are reduced. It is likely that small units will be favoured, because of the difficulty of disposing of the large volumes of waste from big units onto small areas of land. Large units are often most able to spend the necessary capital to control emissions by such methods as slurry injection, separation, etc. Farmers must be ready to take action to control the emissions of substances that are known to be noxious, and should be aware when scientists discover new threats to the environment of intensive farming practices.

References

de Roest, K. (1997) Economic and technical aspects of the relationship between product quality and regional specific conditions within the Parmigiano–Reggiano livestock farming system. In: Sørensen, J.T. (ed.) *Livestock Farming Systems – More Than Food Production*. European Association of Animal Production Publication No 89, Wageningen Pers., Wageningen, pp. 164–176.

Kristensen, E.S. and Halberg, N. (1997) A systems approach for assessing sustainability on livestock farms. In: Sørensen, J.T. (ed.), *Livestock Farming Systems – More Than Food Production*. European Association of Animal Production Publication No 89, Wageningen Pers., Wageningen, pp. 16–29.

van Bruchem, J., van Os, M., Viets, T.C. and van Heulen, H. (1997) Towards environmentally balanced grassland-based dairy farming – new perspectives using an integrated approach. In: Sørensen, J.T. (ed.), *Livestock Farming Systems – More Than Food Production.* European Association of Animal Production Publication No 89, Wageningen Pers., Wageningen, pp. 301–306.

Wilkinson, J.M. (1981) Potential changes in efficiency of grass and forage conservation. In: Jollans, J.L. (ed.) *Grassland in the British Economy.* Centre for Agricultural Strategy, Paper 10, University of Reading, Reading, Chapter 26, pp. 414–429.

Further Reading

Ap Dewi, I., Axford, R.F.E., Marai, I.F.M. and Omed, H. (eds) (1994) *Pollution in Livestock Production Systems.* CAB International, Wallingford.

Gasser, J.K.R. (1980) *Effluents from Livestock.* Applied Science, London.

Jones, J.G. (1993) *Agriculture and the Environment.* Ellis Horwood Series in Enviromental Management, Science and Technology, Chichester.

Phillips, C.J.C. and Piggins, D. (eds) (1992) *Farm Animals and the Environment.* CAB International, Wallingford.

Sørensen, J.T. (ed.) (1997) *Livestock Farming Systems – More Than Food Production.* European Association of Animal Production Publication No 89, Wageningen Pers., Wageningen.

Taiganides, E.P. (1977) *Animal Wastes.* Applied Science, London.

The Future Role and Practice of Cattle Farming 10

Introduction

Cattle farming is a long-term commitment, and as such is usually regarded as an inelastic business. Cattle farmers cannot react to changing demand rapidly because of long production cycles and the high level of investment required to provide the necessary resources, especially for milk production. Occasionally, cattle production systems have had to change rapidly, such as in former Eastern Bloc countries, or as a result of new diseases, such as following the outbreak of bovine spongiform encephalopathy (BSE) in the UK, which had major effects on consumer confidence in the safety of cattle products and hence demand. Such instances can result in instability in cattle production systems, with businesses failing and sometimes even the social structure of the region being affected. More gradual change results from technical developments and changing dietary habits. Reliable, long-term forecasts of the requirements for cattle products will enable changes to be made most efficiently. This chapter describes the changes that are taking place in these requirements and the effect that this is having on the way in which cattle businesses are managed in the rural environment.

The Relationship between Cattle Production Systems and the Environment

As the world population of farm animals increases to cater for the increased human population, their effect on, and interaction with, the environment becomes of major significance. This has stimulated public concern for improved cattle production systems, often also called sustainable, where the environment is not adversely affected in the long or short term, the products are safe for the consumers and the animals have a high quality of life.

If such systems are not economically viable, particularly if labour costs are high, external support may be provided in the form of government grants. The direction of these grants will depend on the public need to maintain different systems of animal production. In regions of the world with a long history of settled agriculture there is a public reluctance to see traditional, extensive systems of cattle farming disappear. Grants which were originally provided to support high production levels are increasingly being diverted from direct support for food products to support for maintaining traditional systems or features of environmental value in the countryside. Support may also be provided to maintain farmer income when natural disasters affect the industry, such as during major outbreaks of disease or collapse in public confidence, and this will preserve stability in the cattle industry. Stability is further enhanced by making changes in support gradually, but sometimes, new governments may bring about rapid changes that damage the industry. Rapid withdrawal of governmental support for dairy farmers in New Zealand in the 1980s demonstrated that it takes several years for farmers to adapt their systems and if necessary find new markets. The period while farmers adapt to new political directives can be destabilizing and lead to long-term damage to the production system that takes years to establish. This happened in the former Eastern Bloc countries, where not only was much of the government support and control rapidly withdrawn, but there was enforced land redistribution, as state and collective farms were returned to their owners before the land was collectivized, which was often at least 50 years ago. Alternatively, the land was distributed between the farm workers. In addition to these difficulties, there was a marked loss of purchasing power for cattle products, which are expensive compared to other staple foods. As a result of escalating costs of animal production and reduced prices for the products, many of the new owners were unable to operate a viable farming system and land has since changed hands several times. An increasing proportion of the land is returned to systems of production requiring fewer resources than cattle farming, such as sheep production.

In the developing regions of the world, expanding populations and increasing demand for cattle products create a need for sustainable and efficient systems of production that use the most appropriate technology. In the latter part of the 20th century, many developing countries relied on surplus milk powder from the subsidized dairy industry in the EU to support their infant feeding programme. Now that the European milk powder surpluses have been brought under control and there is a growing urban population and therefore demand for milk products, many developing countries are expanding their dairy farming industry. The new dairy systems are often based in the periurban districts, with the major difficulty being lack of suitable fodder for the cattle. It would be preferable for long-term sustainability of the human population to contain dairy production to the marginal land, where cattle fodder can be grown easily. Other land can then be maintained for the production of more demanding crops – food crops, fuel (e.g. oil crops) and raw materials (e.g. fibre crops). Additional land may be kept in its indigenous vegetation to attract tourists and to maintain a gene pool of diverse biological material. These multiple land-use systems are evolving as a result of market forces in some less developed regions of the world, where government support for the land-based industries has traditionally been limited. It might evolve faster under the *combined* influence of market forces and strategic support from more developed regions to prevent environmental damage and over-exploitation of natural resources, but the latter is in short supply, as the industrial nations do not always take a sufficiently long-term view of the benefits of collaborating with developing nations.

In future, land for cattle production is likely to see increasing competition from other potential users, because cattle products can be replaced by more efficient forms of food production, and it is likely that sustainable forms of fuel and raw material production will require more land in future. Cattle have been an easy means of producing food from land, but as land becomes scarcer, their efficiency will be increasingly challenged by alternative land users.

The Future Market for Cattle Products

No consideration of the future role of cattle in our countryside is complete without consideration of the future demand for cattle products, as this will largely determine the economic climate in which farmers will operate. This must be put in the context of a world population living in either industrialized regions that have low or zero population growth, or increasingly destitute countries that are trying to develop, but are hindered by rapidly expanding populations, debt-repayment obligations and loss of market potential as a result of the industrialized countries protecting their market. The extent of freedom of trade is of vital importance to the future demand for cattle products in both the developing and developed world. Tariffs restrict the access of the developing countries to markets in the developed world, where the demand for high-priced cattle products is strong. However, greater trade would be possible if these tariffs are illegitimized, and the world could benefit from the technological developments that allow cattle products to be safely stored and transported. Many tariffs are introduced on safety and increasingly on welfare grounds, but increasingly the developing countries have the technology to produce to the highest standards, if required. The situation will become more acute as developing countries increase their share of the total world population, Africa for example will double its share of the worlds' population by 2052 (from 12% to 24%). The former Eastern Bloc countries are trying to recreate a western-style market economy, but this will increase the rich/poor divide and may take the responsibility for environmental protection out of central control. The rapid reduction in profitability of dairy farming in these countries and the dissolution of many of the large state and cooperative farms is producing a return to extensive farming methods, at least temporarily until the new land owners can accrue the capital to invest in their businesses. Extensive grazing systems are beginning to be used again and, in some cases, tractors are being replaced by animal traction.

History will no doubt remind us at some later date that the development of a sustainable cattle farming industry may benefit from a certain degree of central support, otherwise the self-preserving nature of the farmers could allow land resources to be squandered for short-term profit. However, any central control has to be for the long-term benefit of the farming system and not to increase the productivity of the land at the expense of its long-term sustainability. Both the 5-year plans of the Communist era and the product support of the Western European governments could be criticized on this account. Former successful agricultural systems developed centrally managed insurance policies, such as resting the land periodically or creating stores of food reserves to guard against the adverse conditions that occasionally afflict farming systems. It has been known for many years that land fertility will decline unless account is taken of the need to return to the soil the resources that are removed by farming. The globalization of other world industries, often with central control by big multinational companies, has so far largely ignored cattle farming, partly as a result of attempts by Western governments to preserve traditional systems. Whilst plausible for limiting the rate of change in a vulnerable sector of the rural population, it is likely that globalization of the cattle industry will happen, and that the degree of central management that this will bring could increase the sustainability of the industry.

The demand for beef

In many industrialized nations, the increased affluence of the last 50 years has stimulated an increase in meat consumption. In Europe, a post-war policy of increasing agricultural production was increasingly successful up to the early 1980s. Coupled with an increase in disposable income, this led to an increase in meat consumption from 50 kg year^{-1} per head in the 1960s to the current average of 87 kg year^{-1} per head. This is still considerably less than consumption in the USA, which averages 115 kg year^{-1} per head. However, recently meat consumption has begun to decline in some Western nations, particularly red meat and particularly in the young members of the population that usually indicate future trends in consumer demand. There are three main reasons for this.

1. Human health concerns: there have been many assertions that eating meat, particularly red meat is detrimental to human health. This is an over-simplification. The overriding worldwide influence of meat consumption is undoubtedly positive, since malnutrition is still widespread in developing countries and meat is rich in available nutrients. Protein consumption is often inadequate, and meat has a high concentration of this nutrient. Iron-deficiency anaemia is also commonplace, and meat is a good source of iron of high availability. It is also a good source of B vitamins.

In the industrialized world, potential risks of meat consumption are because of the excessive nutrient intake of a large proportion of the population.

Increasingly sedentary lifestyles in Western countries have reduced nutrient requirements. However, epidemiological studies have demonstrated that the consumption of meat, and the high blood-cholesterol levels that tend to follow, are not necessarily bad for human health, indeed the incidence of gastro-intestinal cancers appears to be slightly increased on low-meat diets. The major risk is that fat consumption on a high-meat diet will be excessive and obesity will ensue, which places increased demands on the cardiovascular system. Meals with high carbohydrate contents lead to a greater feeling of satiation than isoenergetic meals with high fat contents. This is particularly true for the structural carbohydrates (fibrous products), and there is a corresponding reduction in appetite. In relation to mineral supply, cereal and other high-fibre products consumed in Western countries are often fortified with added minerals and vitamins to ensure adequate intake, although deficiency problems are more often seen in developing countries as total intakes are low.

2. Animal welfare: the intensification of cattle production systems has height-ened awareness of the welfare of the animals, with many people citing animal cruelty as the main reason for non-consumption or low consumption of meat.

Intensification is often believed to lead to a reduction in animal welfare. This is the public perception of modern animal production systems, but we have few objective criteria with which to judge the welfare of our farm livestock as yet, and many of the fundamental questions remain unanswered and often unconsidered. Do cattle prefer a short, happy life to a long one in poor conditions? What are their needs for mental stimulation and how do these interact with physical requirements, of which we have a better knowledge? Is an anthropomorphic evaluation of cattle needs, so often used by the public to judge production systems, a useful guide or an unnecessary irrelevance? How does the animal's perception of its wellbeing change during its lifetime?

In the EU, intensification has been most evident in the pig and poultry industries, yet it is the consumption of these meats that has been maintained, while that of beef and lamb is declining in some countries. This anomaly suggests that other factors than welfare concern are also influential for changing meat consumption habits. Relative cost is one of these, with the beef and sheep industries failing to reduce relative prices through the use of modern technology, as the white meat industries have done.

3. Competition from vegetable-based foods: that meat has been a natural food for humans throughout their evolution is undeniable. However, the rapid development of the food industry, particularly in industrialized regions of the world, has produced increased competition from other staple foods, with meat often appearing *relatively* less attractive. The food manufacturers have developed non-meat foods that appeal strongly to all our senses. Consider, for example, the growth of the breakfast cereal market, which has replaced the traditional egg and bacon cooked breakfast in the UK. Cereals are ultra heat-treated to improve digestibility and coated with a large variety of sweet, nutty or aromatic substances to stimulate the gustatory senses. The food-processing industry has largely concentrated its efforts on non-meat foods, because meat with its high raw material cost has less potential than, for example, cereals for added value. The visual appeal of non-meat foods has also been exploited to the full, and the full range of colours of breakfast cereals, for example, is instantly attractive to a child's visual palate. By contrast, the visual and gustatory attractions of meat are increasingly less obvious to many consumers. The appeal of strong-flavoured meat, with its complex volatile flavour compounds, is probably acquired during childhood. It is likely that children are conditioned to enjoy the taste of strong-flavoured meat, in the same way as the enjoyment of spiced, mouldy and smoked foods can be learnt. It is even possible that animals developed the pheromones that produce volatile meat flavours partly to prevent them from being eaten – a prey animal's equivalent of plant toxins. It would be surprising if animals had not developed such adverse flavour compounds during the long course of predator/prey evolution. The experience of zoo-keepers suggests that the consumption of red meats that are highly flavoured, such as the meat of male goats, is an acquired taste for many predators.

These are some of the issues that explain the decline in red meat consumption, particularly by young people in developed countries. In developing countries, the increase in population is often leading to a reduced availability of meat per person. Even if people have no ethical reason to refrain from meat eating, it is highly likely that red meat consumption will continue to decline in the face of increasing competition from plant-derived foods that have been flavour-enhanced or modified in other ways. There is already an increased demand for meat with low fat contents or mainly fats that are protective against heart and circulatory diseases, such as the omega-3 fatty acids. The manner in which many beef cattle are raised, with high-energy diets and lack of exercise, leads to rapid rates of fattening, with the deposition of large quantities of saturated fat in the muscle tissue. In future, discerning meat consumers may require an extensification of meat production systems, with a return to grazing systems, and they may pay more for the products, which will have smaller amounts of intramuscular fat, that is less saturated. Such a product would also satisfy the demand for the animals to be raised in high welfare conditions, as it would be viewed as more natural than indoor fattening.

The vegetarian viewpoint

Most vegetarians do not accept the consumption of other animals on ethical grounds, ostensibly because it demonstrates man's dominion over them. For these people the concept of 'exploiting' captive animals for meat production is unacceptable. The human carnivore may justify his or her actions by claiming that without farming systems farm animals would have no life at all, but this belief does not conform to modern ideals of human and animal welfare. We do not yet know enough about the relative importance of the loss of certain freedoms to determine accurately whether the life of cattle is satisfactory to most consumers. However, the systems are most often criticized for not offering cattle basic resources that we as humans would value for our mental and physical health, such as adequate space, companionship, 'natural' surroundings, perhaps reflecting our highly complex social requirements. In relation to space, we are all captive to a certain extent, humans and cattle, in the biological system in which we function. For humans, this may involve spending most daylight hours in an office, or for farm livestock the stable. We all function in a hierarchical structure, which is the basis of a complex society, and we welcome the existence of distinct territorial boundaries (personal space) as increasing our security. Some, but not all, cattle suffer stress in close confinement and develop behavioural modifications (e.g. stereotypies) to help them to cope. It is difficult for us to criticize cattle production systems for having inadequate space per animal until we know precisely the requirements that cattle have and their tolerance of space allowances that differ from the optimum.

The future demand for milk and milk products

Milk and milk products were not a natural part of the adult human's diet until cattle were domesticated 8000–10,000 years ago. They are naturally high in fats, as a proportion of their total solids, which will tend to lead to obesity when consumed in large quantities. However, most humans are conditioned to accept cows' milk and milk products as infants, and probably will continue to be, given the difficulty in replacing milk with vegetable-based products that can be adequately digested by babies. However, the major factor governing the maintenance of demand for milk products is the adaptability of the raw material, as it is with cereals. The food processing industry manipulates the milk constituents in ingenious ways to produce palatable and convenient foods, utilizing all the time our conditioned attraction to dairy fats. As with meat, only small inclusion rates of the fat are actually necessary to impart the necessary flavours, which has led to a profusion of mixed dairy/vegetable fat products, or low-fat dairy products. These may be imparted with sufficient viscosity to improve handling properties by artificial thickening agents. There seems little reason therefore to be pessimistic about continued demand for dairy products, albeit of reduced fat concentration in the industrialized countries. Doubts about the welfare of dairy cattle kept in intensive units may strengthen, and again more research is needed to understand their needs. Intensive production and a high standard of animal welfare are not irreconcilable, but there is a problem of human perception in the absence of scientific facts.

The Role of Cattle in a Multiple-purpose Land-use Context

From the previous discussion, it is evident that the role of cattle will not simply be to provide products for human consumption in future. In some situations, they can be regarded as stewards of the countryside and must meet the public needs for high welfare systems that maximize environmental protection. The demand for extensification in industrial countries, and the need to confine cattle to marginal land or as complements to arable enterprises in developing countries, will inevitably change the systems in operation and farmers will need to be flexible to survive.

In many areas, tourism will assume a greater role in dictating land use, even if it is in combination with cattle production systems. In the UK for example, tourism employs over half a million people, equivalent to the total number of people employed in agriculture, but the proportion of the total land area mainly dedicated to tourism is very small.

The future role of cattle production systems may be summarized as:

1. To provide food, fibre and fuel products that maintain and contribute to the health of the human users.
2. To provide conditions for cattle that meet their physical and mental needs, as perceived by the general public and in particular the consumer.
3. To preserve and foster a countryside that is of high biological and aesthetic value to the human population. The biological value includes the maintenance of diverse flora and fauna, and maintaining natural resources such as farm woodlands.
4. To minimize pollution from cattle farms that could damage either the microenvironment of the farm itself, its surroundings or the macro or global environment. Particular difficulties exist with liquid and gaseous effluents from intensive units.
5. To coexist with and complement alternative land-use systems, e.g. forestry (as in silvopastoral systems), arable production (where by-products may be usefully converted into a product that is of value to humans) or tourism.
6. To preserve the biodiversity of the cattle population, in particular to enable cattle to fulfil a useful function in future, when requirements may change or new disease challenges require certain genotypes within the cattle population.

Additional possible minor roles for cattle are the production of pharmaceuticals in milk or blood by transgenic manipulation, and limited use for sport, although sports which are incompatible with (**2**) above are increasingly unlikely to be accepted by the public.

There must be adequate provision for a high welfare environment, especially in the short term by manipulating the environment to suit cattle, but in the long term this may include genetically manipulating the animals to suit the environment. However, it is in our interest to maintain a reasonably intact gene pool for future insurance.

Genetic insurance

Theoretically, in today's age of rapid genetic manipulation techniques, our cattle could be transformed into meat and milk producing 'vegetables', incapable of normal behaviour and with gross distortions of body morphology. To some people, cattle breeding has already gone too far in this respect by, for example, producing cattle breeds with muscular hypertrophy that can result in a high proportion of dystokia during calving unless

Caesarean operations are routinely performed. It would be an irony if at the very time when we are trying to maintain genetic diversity in our wild flora and fauna, we deny cattle their genetic inheritance and diversity. This is their security for future generations, which must be the main priority of all species, wild or domesticated. In this time of very rapidly changing agriculture and countryside management it would be unethical to deny cattle their genetic inheritance necessary for long-term survival. There are three main reasons why the maintenance of genetic diversity in our cattle is important:

1. Loss of environmental adaptability: the rate at which an organism can adapt to meet changing environmental circumstances is dependent on the diversity of its genotype. For example, animals in the *Bos* genus originally had a limited role as grazing animals in Asia, and following domestication the species diversified into a wide range of genotypes to adapt animals to varying environmental conditions in which it was kept. In recent years, some of this genetic diversity has been lost, as the Holstein–Friesian has become the dominant genotype for intensive milk production systems. The cows of this breed require food of high nutrient density, they are more susceptible to hot conditions and often have less disease resistance than cows with lower production potential. Their milk is of low solids content and is therefore relatively expensive to transport and process. In the future, resistance to adverse environmental conditions and low milk transport and processing costs may be more important than a high milk yield per cow, the major benefit of which is to reduce the associated labour requirement. The genotypic information needed to adapt to new demands must be preserved, if production is to be efficient in relation to future resources.
2. Human security: man's manipulation of the environment is far less impressive than the product of millions of years of evolutionary development. Humans need to be reassured that their future compatibility with the natural environment is secure, and this is unlikely to be the case unless rural biodiversity is preserved. Increasingly people need to complement a stressful working environment, which is often in artificial surroundings, with relaxation in a countryside which contains evidence of natural variation and sustainability, such as exists in some cattle production systems.
3. Product diversity: people's dietary habits change and the need to allow for changes in human dietary requirements necessitates the maintenance of product diversity. Variation in food type is also part of cultural identity and a nation's heritage, without which life would be less unique to individuals.

The maintenance of biodiversity in natural fauna and flora on cattle farms
Man has to provide a range of grazing pressures on grassland to create a diverse environment. Further research is needed on the effect of cattle management practices on the flora and fauna in the countryside, but we should not necessarily assume that our dominant farm herbivores, cattle, always create the best environment for our natural flora and fauna. Cattle, although not naturally browsers, can do considerable damage to shrubs and trees in rangeland where there is insufficient grass, as can natural browsers

such as goats and deer. However, they are unselective grazers, which makes them suitable to maintain a wide variety of pasture species, whereas selective grazers such as sheep can preferentially graze some plant species and reduce their competitiveness.

The main necessities for maintaining a diverse flora, and hence wildlife, are to maintain a diverse range of stocking densities and not to overstock the pastures. Wild or range animals have often been depleted in numbers to make way for single-purpose cattle production systems at the expense of the environment and biodiversity of the region.

Diverse flora and fauna can be accomplished by maintaining family (matriarchal) groups of cattle, as the animals within a group have different foraging strategies according to their individual needs (or physiological state) and morphology, particularly the shape and size of their mouth. More effective, however, is the mixed or rotational grazing of different livestock species with cattle. Cattle and sheep are often grazed together but do not have very dissimilar grazing horizons. Cattle are more complementary to the feeding habits of browsers, such as goats or deer. Sheep are not only selective in the herbage species they select, they are able with their small mouthparts to select only the young leafy vegetation and leave the old, brown stem. This makes a sward unsightly and unproductive but it does provide a residue of herbage that the cattle will eat if there is nothing else available (foggage). Foggage used to be used in traditional farming as a standing hay crop for winter fodder on free-draining farms. The practice encourages the more erect grass species such as Yorkshire fog (*Holcus lanatus* L.) and cocksfoot (*Dactylis glomerata* L.) and reduces the white clover (*T. repens* L.) content. Cocksfoot is particularly prone to winter frost damage. In the long term, 'fogging' can open up the sward to invasion by novel species and increased biodiversity. If old mature herbage is thought to be unsightly or wasteful, grazing sheep pastures with cattle will remove much of the dead material. Cattle can be of similar benefit to horse-grazed pastures.

Silvopastoral systems will delay the maturation of herbage but this will also reduce the seed set by plants, resulting in some loss of annuals . A system of cattle grazing amongst fruit trees has been employed effectively in many temperate regions for centuries, producing high-quality pasture for the animals, fertilizer for pasture and trees, and amelioration of the environment for the animals, all of which are more difficult and costly to provide in dedicated single-purpose systems. The value of such a system is obvious, with a high regard for the welfare of the cattle, species diversity and a variety of cattle and plant products, that indicates increased self-sufficiency and economic insurance for the farmer and his family. Most modern silvopastoral systems use trees for timber rather than fruit, since the fruit production industry has not yet come under the same sort of pressure for extensification as other agricultural sectors. Also, there continues to be a strong demand for timber worldwide.

The timing of grazing will also have a distinct impact on vegetation diversity and composition. Traditional hay meadows with their varied flora can only be maintained by a precise management regime, which has long since been abandoned by most 'output-oriented' farmers. Cattle are over-wintered on straw, with the resulting farmyard manure being spread in the spring for a limited but prolonged nutrient release to the pasture. Meadows are grazed by cattle or sheep up until mid-May and then rested until a late hay cut in July/August, by which time all the annual flowers have produced their seeds. Subsequently the meadows are lightly grazed by cattle and sheep in the autumn and sheep only in winter. Similar management strategies are available for maintaining other scarce or diminishing systems such as grassland on calcareous soils, water meadows and heather moorland. If the intention is to restore species diversity to ancient meadows, it should initially be cut early, perhaps even in late spring, to prevent seeding of the pasture. Seed of the desired species may then be introduced artificially and the floral diversity subsequently maintained by late or no cutting. Early cutting is damaging to the nesting sites of ground-nesting birds.

The Role of Cattle in Developing Countries

In developing countries, the role of cattle will be by necessity different. Governmental or public support is unlikely to be available for preserving traditional cattle production systems (if they exist) in the way previously described for the UK. There can be little moral justification for expanding the area under cultivation for cattle feed to satisfy the needs of a small minority for meat or foreign capital, while food supplies are inadequate, and inadequate fuel, for example for cooking, may be causing considerable deforestation. The rising population inevitably will mean greater areas under cultivation and perhaps even greater pressure than before to confine cattle production to the marginal areas and to the use of industrial and other by-products. There is an inevitability about increasing areas under cultivation in regions where the human population is increasing, and this will reduce the extent of indigenous landscapes.

The role of cattle in arresting environmental degradation

Well-planned cattle production systems can play a useful role in both developed and developing countries in arresting environmental degradation. Silvopastoral systems with cattle are being used to arrest soil loss and encroaching desert in many areas. Dung from the animals adds useful organic matter to the soil and trees provide shelter for the animals, stability to the soil and aid water and nutrient cycling. The importance of recycling as much as possible within a cattle farming system is self-evident, but is often practised more extensively in developing than developed countries.

Conclusions

Systems of cattle production are changing in response to new economic and political pressures and changing moral values in the population today. The recent intensification of cattle production worldwide is being reversed in many developed countries because of public concern for the effects on the environment and animal welfare, and in many Eastern Bloc countries because of the dissolution of large state farms and a lack of capital to finance intensive milk production systems. The increase in the consumption of meat in Western countries in recent decades is now being reversed in many regions because of concerns over the ethics of intensive animal production, health concerns and increased competition from other, highly processed foods. The consumption of milk and milk products is unlikely to decline because of their adaptability and palatability, but health concerns will require greater production of low-fat milk.

Cattle will in future have to play a greater role in complementing and supporting non-food enterprises in the rural environment, particularly tourism and fuel and fibre production. Methods of managing cattle must ensure the maintenance of biodiversity in the rural environment, both in respect of indigenous flora and fauna, and in the maintenance of a diverse gene pool in the cattle species to secure their future use in farming or natural habitats.

Further Reading

Jones, J.G. (1993) *Agriculture and the Environment.* Ellis Horwood Series in Enviromental Management, Science and Technology, Chichester.

Kyle, R. (1987) *A Feast in the Wild.* Kudu, Oxford.

Payne, W.J.A. (1985) A review of the possibilities for integrating cattle and tree crop production systems in the tropics. *Forest Ecology and Management* 12, 1–36.

Phillips, C.J.C. and Piggins, D. (eds) (1992) *Farm Animals and the Environment.* CAB International, Wallingford.

Phillips, C.J.C. and Sørensen, J.T. (1993) Sustainability in cattle production systems. *Journal of Agricultural and Environmental Ethics* 6, 61–73.

Index

This book is due for return on or before the last date shown below.

Don Gresswell Ltd., London, N21   Cat. No. 1207
DG 02242/7